William E. Merrill

Iron Truss Bridges for Railroads

Methods of calculating strains, with a comparison of the most prominent truss bridges, and new formulas for bridge computations

William E. Merrill

Iron Truss Bridges for Railroads

Methods of calculating strains, with a comparison of the most prominent truss bridges, and new formulas for bridge computations

ISBN/EAN: 9783337350680

Printed in Europe, USA, Canada, Australia, Japan

Cover: Foto ©berggeist007 / pixelio.de

More available books at **www.hansebooks.com**

IRON TRUSS BRIDGES FOR RAILROADS.

METHODS OF CALCULATING STRAINS,

WITH A COMPARISON OF

THE MOST PROMINENT TRUSS BRIDGES,

AND NEW FORMULAS FOR

BRIDGE COMPUTATIONS;

ALSO,

THE ECONOMICAL ANGLES FOR STRUTS AND TIES.

BY

BVT. COL. WM. E. MERRILL, U. S. A.,
MAJOR CORPS OF ENGINEERS.

ILLUSTRATED WITH NINE LITHOGRAPHIC PLATES.

SECOND EDITION.

NEW YORK:
D. VAN NOSTRAND, PUBLISHER,
23 Murray Street and 27 Warren Street (up stairs).
1870.

Entered according to Act of Congress, in the year 1869, by

D. VAN NOSTRAND,

in the Clerk's Office of the District Court of the United States for the Southern District of New York.

PREFACE.

HAVING been accidentally put in the way of studying truss bridges, and having at first labored under great difficulties from inability to procure clear and simple instructions how to calculate strains in trusses, being compelled to work upward from the simplest combinations, the author of this work has thought that the results of his labors might be found useful for others striving after the same object. Finding that the usual formulas for the strength of pillars were not in suitable shape for use in bridge calculation, they were remodelled to suit all cases in practice, assuming Hodgkinson's formulas to be reliable. To determine the exact value of these formulas, much time was devoted to an analysis of the experiments on which they are based, with the results given in the text.

In order to ascertain the best form of combination in trusses, a very close calculation was made on the seven best-known forms of truss in use in this country, every care being taken to subject all to identically the same conditions. After finishing these calculations an effort was made to ascertain the economical angles for struts and ties, and from them to determine the theoretically best combination, and the reasons for the observed differences in the trusses examined.

All the steps in each investigation are given in full, that every one may judge for himself of each step. It is hoped that even those who may differ from the conclusions in reference to trusses, will yet find the work valuable for its formulas, which, as strict deductions from Hodgkinson, are entitled to receive the same credit that is universally accorded to the originals. The author will be fully satisfied if his efforts to solve the problems in question should be of service in guiding others to their complete solution.

W. E. M.

CONTENTS.

	PAGE
GENERAL INTRODUCTION	7
METHOD OF CALCULATING STRAINS IN TRUSSES	11
HODGKINSON'S EXPERIMENTS ON CAST-IRON PILLARS	25
FORMULAS FOR OBTAINING THE VOLUME AND WEIGHT OF A WROUGHT-IRON TIE UNDER A TENSILE STRAIN	36
FORMULAS FOR OBTAINING THE STRENGTH, VOLUME, AND WEIGHT OF A CAST-IRON PILLAR UNDER A STRAIN OF COMPRESSION	37
CONSTANTS USED IN CALCULATING WEIGHTS	51
THE FINK TRUSS	53
THE BOLLMAN TRUSS	62
THE JONES TRUSS	68
THE MURPHY-WHIPPLE TRUSS	78
THE POST TRUSS	85
THE TRIANGULAR TRUSS	93
THE LINVILLE TRUSS	103
COMPARISON OF WEIGHTS OF BRIDGES	110
BEST ANGLE FOR A PAIR OF TIES	113
BEST ANGLE FOR A SET OF TIES	115
BEST ANGLE FOR A PAIR OF STRUTS	117
BEST ANGLE FOR A SET OF STRUTS	121
LATTICE BRIDGES	124
USE OF WROUGHT-IRON IN COMPRESSION MEMBERS	126
CONCLUSION	127

LIST OF ILLUSTRATIONS.

	PAGE
I., II. HODGKINSON'S EXPERIMENTS ON CAST-IRON PILLARS	25
III. THE FINK TRUSS	53
IV. THE BOLLMAN TRUSS	62
V. THE JONES TRUSS	68
VI. THE MURPHY-WHIPPLE TRUSS	78
VII. THE POST TRUSS	85
VIII. THE TRIANGULAR TRUSS	93
IX. THE LINVILLE TRUSS	103

IRON TRUSS BRIDGES FOR RAILROADS.

THE railroad interest of the United States has become of late years one of the greatest in the country, and every day witnesses a steady increase. All matters connected with our railroad system concern every citizen, whether a stockholder or not, as every kind of business, or even of pleasure, is more or less dependent upon the safe and rapid interchange of persons and things that has sprung up from, and is the natural consequence of, railroads. The enormous amount of capital invested in them, and the inestimable value of the human lives daily intrusted to their care, make everything tending to promote true economy and safety in their management a matter of the greatest importance to all.

Of all parts of a railroad the bridges are the most expensive, demand the most skill in construction, and cause the greatest loss of life in case of failure. Every effort to ascertain with mathematical certainty the true strains on every part of a bridge, the proper manner of proportioning each part, and the best form of combination to secure the greatest economy with the needed safety, should receive the hearty welcome of the public. There are none of us who do not almost daily risk our own lives, or the lives of those dear to us, on the skill and fidelity of the men who build our railroad bridges. And yet it is lamentable to recall the number of accidents from the failure of bridges, that are within our memories.

The object of the present paper is to endeavor to throw additional light upon the method of calculating the maxima strains that can come upon any part of a bridge truss, and upon the manner of proportioning each part so that it shall be as strong relatively to its own strains as any other part, and so that the entire bridge may be strong enough to sustain several times as great strains as the greatest that come upon it in actual use. This multiplier of the actual strains we call the Factor of Safety, and it is the number by which we multiply all the calculated strains to get the strains which we actually use for proportioning the different parts of the truss. We will assume 5 as the Factor of Safety for a Railroad Bridge, and we will so proportion all the parts of each truss that they will just give way under these augmented strains. This may seem an excess of precaution, but we have to allow a wide margin for defects in manufacture, and for imperfect knowledge of the laws of the strengths of our material; and, moreover, the value of the lives that daily pass over our railroad bridges is so immeasurable that we must demand safety without regard to increase of cost.

In the early construction of railroads in this country it became necessary to use the most economical plans and materials to be able to build them at all. For this reason all of our first railroad bridges were constructed of wood, wherever the span did not permit the use of brick or stone. But the life of a wooden bridge is so brief that with their increased resources our railroads now find it economical to replace perishable wooden bridges by iron ones that will practically last forever, and thereby be cheaper in the long run. As iron is an expensive material, and moreover one that can be put into the exact shape that will develop its greatest strength, and as every ounce of weight in the bridge itself that is not absolutely required is not only a loss but an injury, it becomes a matter of very great importance to determine the best combination for securing the strength required with the minimum material, and therefore at the least cost.

There are many kinds of trusses competing for public favor, and we will examine the seven best known in this country, and endeavor to ascertain which requires the least material with the required strength, and the reason for its superiority. The trusses which we will examine are known from the names of their builders and designers, or from the character of the combination, as

THE FINK,
THE BOLLMAN,
THE JONES, OR HOWE,
THE MURPHY-WHIPPLE, OR REVERSED HOWE,
THE POST,
THE TRIANGULAR,
THE LINVILLE, OR PRATT.

We may here premise that safe bridges can be built on any one of these plans, but that some require more metal than others to secure the same strength. To carry out our design, we will complete the calculations for each bridge, assuming the same span, the same panel length, the same depth (where the arrangement will permit), the same weight of train, the same assumed weight of bridge, and the same Factor of Safety. If the calculations are not correct, they will at least have the merit of being given in detail, so that any one may readily discover errors for himself.

The effect of competition is such that many bridge-builders, while professing to use 5 or 6 as their Factor of Safety, really use a smaller number. It is to be hoped that every railroad engineer will so familiarize himself with the theory of bridge construction as to detect and prevent any such frauds on the companies and on the public. In no branch of manufacture are skill and honesty more requisite.

Assumed Weights and Dimensions.

Span......................... = 200'
Number of panels .... = 16
Length of a panel .... = 12'0"
Weight of engine............. = 88000 lbs.
" tender............. = 50000 "
" cars............... = 2104 " per foot of track.
" bridge............. = 300000 "

We will suppose each bridge to be of the class termed "through" or "overgrade," composed of two trusses and the necessary flooring and top bracing. This is the form of bridge that must be used in most cases, and it is therefore the proper one to adopt in a comparison.

We will discuss one truss at a time, and therefore we must halve all the weights assumed above. We therefore have for one truss—

Weight of engine............. = 44000 lbs.
" tender............. = 25000 "
" cars............... = 1052 " per foot of track.
" bridge............. = 150000 "

And on each panel—

Panel weight of engine....... = 17600 = $w' + e$
" " tender....... = 16160 = $w' + t$
" " cars......... = 13152 = w'
" " bridge....... = 9375 = w''
Excess of engine over cars.... = 4448 = e
" tender " .... = 3008 = t

The weight of the bridge itself we will consider as a number of weights of 9375 lbs. suspended at each roadway bearer. This will be sufficiently accurate for our purpose of comparison, though, after the approximate weights of all the parts are obtained by this method, the calculations should be re-made with the bridge weight as it really is. We will find, further on, that the assumed weight of bridge is in excess of the actual weight in all but one of the trusses examined. It is more convenient for the purposes of comparison to have it in excess than in deficiency.

Nomenclature.

It will conduce to clearness and exactness to use the same nomenclature for all the bridges discussed in this treatise. The top of a truss is variously termed the Boom, the Top Chord, or, simply, the Top. The bottom is called the Chord, the Bottom Chord, or the Bottom. Strictly speaking, the term Chord should be restricted to a member under tension connecting the springing lines of arched bridges; but the terms Top Chord and Bottom Chord are convenient, readily understood, and they are in common use by the largest wooden-bridge-building companies—those that build the Howe bridge. We will, therefore, adopt them, and we will call the top of a truss the Top Chord, and the bottom the Bottom Chord. All diagonal members will be called by the general name of Braces. If they undergo strains of compression, and transfer the weight to the nearest abutment, they will be called Struts; if of tension, Ties. Counter-struts and Counter-ties are those that carry part of the moving load to the more distant abutment. In the same half of a truss, struts and counter-struts, or ties and counter-ties, have opposite inclinations. The distinction will be better understood as the discussion of each form of truss progresses. Vertical, or nearly vertical, members under compression will be called Posts; under tension, Vertical Ties.

The different arrangements for connecting bridges with the abutments, and the different methods of counteracting the disturbing influences of changes of temperature, are not essential to the purpose of this treatise. The devices are generally such as might as readily be used for one bridge as for another, patent rights alone interfering. Such details would be an unnecessary complication, and they are foreign to the use of the truss as a combination for the transmission of strains. It is sufficient to say that all of these bridges have the necessary compensations for thermometric variations. We design comparing the trusses in question with an eye only to *economy in form*—to see which method of combining the chords and braces enables us to carry a given weight with the least material.

In the drawings of the trusses all cast-iron parts will be black, and wrought-iron parts red. As a general rule, counter-braces, whether struts or ties, will be represented by dotted lines of the proper color, in order to be able to distinguish them more readily.

METHOD OF CALCULATING STRAINS IN TRUSSES.

Before attempting to calculate the strains on the different parts of a truss bridge, and, as a necessary consequence, their sizes that they may safely bear these strains, it is necessary to determine the manner in which strains are transmitted from any part of a truss to the abutments on which it rests. The problem in all trusses is to resolve vertical forces acting at points with no supports under them, into vertical forces acting at points that are directly supported by piers or abutments, and, at the same time, to cause the opposite horizontal forces developed in the process of transmission to the fixed points to neutralize each other by opposition through a horizontal member.

Though much discussion has been given to *arches* of all kinds, but few engineers have taken the trouble to work out thoroughly and carefully the problem of *truss* bridges, a problem of great and daily increasing importance to the great railroad interest of the world.

We will endeavor to settle the question of the method in which strains are transmitted to the abutments in truss bridges, and the method of calculating all the parts of a truss bridge; and we will make a comparison between the seven truss bridges selected, endeavoring to ascertain the form of an iron truss bridge which will carry a given weight of train with the minimum amount of metal.

In Figure 1, let $ABDC$ be a truss. The weight W at D, represented in intensity and direction by Da, is at once transmitted to A, where it becomes Ab. There, by the principle of the parallelogram of forces, it is necessarily resolved into Ac, in the direction of AB, and Ad, in the direction of AC. At B, Ac becomes, in direction and intensity, Be; which is at once resolved into Bf, acting vertically downwards and neutralized by the resistance of the abutment; and Bg, a horizontal force acting to cause the tie-beam BC to move to the left. Following Ad, which at C is Ch, we find it resolved into Ci, neutralized by the other abutment, and Ck, a horizontal force tending to move the tie-beam BC to the right. The two horizontal forces Bg and Ck are equal, and neutralize each other through the tie-beam BC, no tendency to move in either direction resulting.

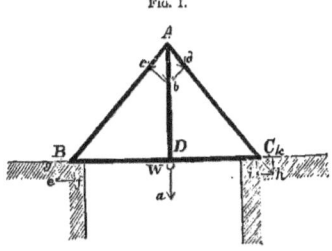

Fig. 1.

It can readily be shown, both by analysis and by graphic construction, that whether

A is over the middle of the tie-beam, or nearer one abutment than the other, the horizontal forces developed are equal, and neutralize each other.

Suppose, as in Figure 2, we have the combination CAB. Let the weight at A be Ad. Drawing the parallelogram of forces we get Af and Ae. $Af = Bg$ is resolved into the vertical force $Bh = Ad$, and the horizontal force $Bi = fd = Ae$. We see from this figure that the horizontal forces developed are again equal and acting in opposite directions, though not directly opposed; that no portion of the weight is transmitted through AC, nor borne by abutment C, and that B supports exactly the entire weight, and no more and no less. We may conclude from this that though a vertical weight may develop horizontal forces in horizontal members, no portion of the weight itself can be carried by them to their abutments, but all the weight must be carried by inclined members, and that if there are no inclined members on the side of one abutment, that abutment will not carry any of the weight, although it will be acted upon by developed horizontal forces.

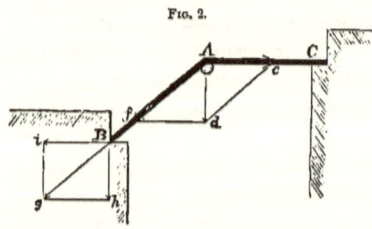

Fig. 2.

If we have but a single beam from one abutment to the other, the proper portions of the weight will be sent to each abutment, and whatever horizontal forces may be developed in the transmission will be neutralized within the beam.

Instead of, as in Figure 2, neutralizing the horizontal force Ae by the abutment C, suppose we introduce an equal and opposite force developed in the same way as Ae by a

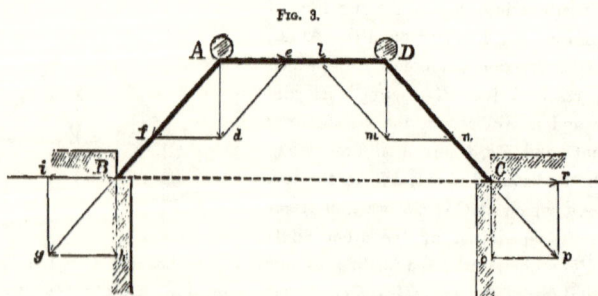

Fig. 3.

weight at D, which is upheld by DC, a strut with the same inclination as AB. The operation of the weight A will in no way be changed, and it will as before go entirely to the abutment B, while the weight at D goes to the abutment C. Their horizontal components $A\,e$ and $D\,l$ meet and balance through $A\,D$.

The horizontal components $B\,i$ and $C\,r$, acting to overturn the two abutments, are equal and opposite. If then a physical connection be made between B and C by a tie along the line $B\,C$, these horizontal components will neutralize each other through this tie, and the abutments will be relieved from any tendency to overturn.

If now we have the combination, shown in full lines in Figure 4, it is manifest that no change in the strains would be produced by supposing the weights at A and D to be at E and F, connected with A and D by ties. The same resolutions and developments of strains take place as before. It is also evident that were the members $E\,D$ and $A\,F$ inserted, their presence would in no wise change the action of the weights, and that no strains would come upon these new members.

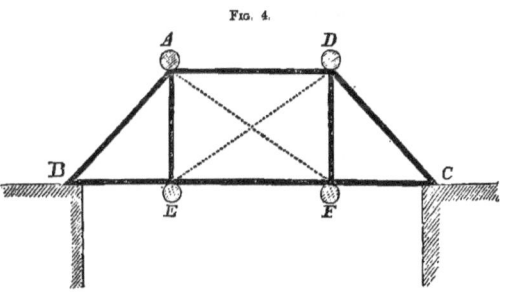

Fig. 4.

If we suppose $A\,D$ lengthened out, as in Figure 5, it is manifest that the action of the weights at A and D must still be the same. The weight at A goes entirely to B

Fig. 5.

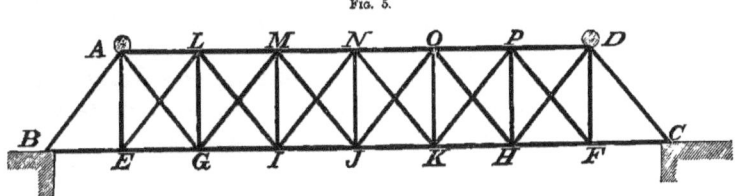

through $A\,B$, developing a horizontal strain through $A\,B$, while the weight at D goes entirely to C, developing a horizontal strain on $D\,A$, equal to and neutralizing that first developed.

If now the weight at A and D be moved respectively to L and P the same kind of resolution of forces continues. One weight goes entirely to B and the other to C, the developed horizontal strains neutralizing each other through $L\,P$. How the weights go to B and C will be shown afterwards. The same thing would take place were one at M and the other at O.

It is a well-known principle of mechanics, that any body acted upon by a system of forces *in equilibrio* is affected by new forces as if the original forces did not exist. If then while the original weights remain at M and O, equal weights be placed at L and P, they will be transmitted to B and C respectively, as if the two first weights were absent, the horizontal forces developed neutralizing each other through LP. Similar action would arise from the placing of additional weights at A and D. The weights at M and O cannot affect any of the inclined members in the rectangles between them, nor those at L and P the inclined members between them.

We see that these weights act in pairs, and also that each pair acts independently of all other pairs, as if they did not exist. If one of the weights of a pair be removed, the other pairs continue to act as before, and the unbalanced weight is transmitted to the abutments, as if it were the only weight on the truss. If these weights were on the Bottom Chord, the Vertical Ties would transmit them directly and unresolved to the Top Chord. and then the resolutions just indicated would take place.

Were Figure 5 a railroad bridge, with a train gradually coming on, we would find that, until it reached the middle of the bridge, all of its weights would be unbalanced. After passing the middle, the train weights would commence to balance (excepting the excess in weight of the locomotive and the tender), until the entire bridge was covered and all of the weights were balanced in pairs. As the head of the train passed off the bridge, the weights on the far end would gradually become unbalanced, until, as the end of the train reached the middle, all of the pairs would be broken and none but unbalanced weights left. Bridges being always symmetrical in reference to the middle point, the bridge weights are always balanced.

The principle explained above is that of *the counterbalancing of equal weights similarly situated on symmetrical trusses*, and is a principle of great importance in the discussion of counter-braces. Each balanced weight goes undivided to the nearest abutment, without producing any strain upon braces lying between it and its counterpart. Without investigation, we might almost infer such action from the well-known axiom, that Nature always works in the simplest and most direct manner.

Transmission of Unbalanced Strains.

Suppose a weight at A equal to Ad, the point A being connected by ties with the points of support B and C.

Forming the parallelogram of forces, we have Af to be transmitted to B, and An

to be transmitted to C. $Af = Bg$ is resolved into Bi, the horizontal force, and Bh, the vertical force. The former tends to overturn the abutment, and the latter to crush it. Similarly the forces at C are Cr and Co. The horizontal forces Bi and Cr, or their equals ln and ef, are equal to each other, because in the parallelogram $Andf$ they are the altitudes of the equal triangles And and Afd, that have the same base, Ad.

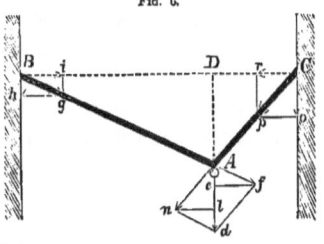

Fig. 6.

We also have

$$Bi : ig :: BD : DA.$$
$$Cr : rp :: CD : DA.$$

The extremes being equal in each proportion, we have

$$ig \times BD = rp \times CD;$$
or, $ig : rp :: CD : BD$.

That is, the portions of the weight transferred to the abutments are inversely as the horizontal distances of the weight from the abutments.

Instead of the very simple combination in Figure 6, suppose the point A, and the weight thereto suspended to be connected with B by the combination $AEBF$. It is manifest that the change makes no difference in the amount and line of direction of the portion of the weight to be transmitted to B. It will as before be Af, and its line of direction will be the theoretical line AB. There being no tie or other means of transmitting the force directly to B, thereby conforming to the general law of mechanics that forces always travel by the most direct route, this force is mechanically resolved into two components, one of which travels through AE, and the other through AF.

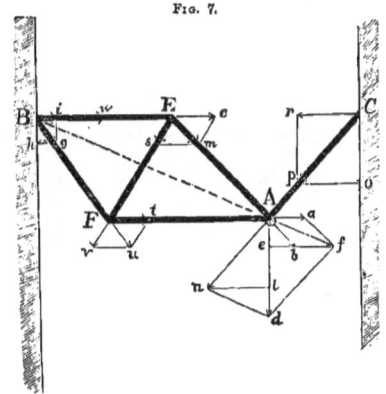

Fig. 7.

The component Ab, that goes through AE, at E, is resolved into $Ec = iw$, that goes directly to the fixed point, and Es that at F becomes Fv, one of the components that generates Fu. The force Aa, transmitted to F where it is Ft, is the other component of Fu, which at B becomes Bg, there being resolved into Bi and Bh. The result then is that Af, the original component of Ad to be transmitted to B, becomes by this mode of transmission Bw acting horizontally, and Bh acting vertically. But Bw, the horizontal force that reaches B, should be equal to ef, the horizontal force to be transmitted, which it is; and Bh should be equal to Ae, which is likewise the case. We may therefore conclude that the transmission of forces has been correctly indicated.

A simple inspection at once gives the strains on all the parts of the combination.

Strains of Extension

on AC is An or Cp
on AE is Ab or Em
on EB is Ec or Iw
on AF is Aa or Ft
on FB is Fu or Bg

Strains of Compression

on EF is Es or Fv

The tendency to pull out the supporting point B is equal to Af, the horizontal component being Bw, and the vertical component Bh. At the point C it is An or Cp, with the horizontal component Cr, and the vertical component Co. Bw and Cr are equal.

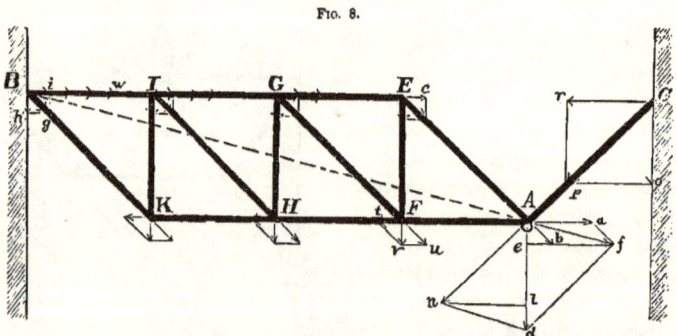

Fig. 8.

In the combination, in Figure 8, it is evident that there is still no change in the proportions of the weight transmitted to the two abutments. The only effect of the combination is to change the details of the transmission. We see that as before Bw, the horizontal component that travels to B, is equal to ef, the amount that should have gone there, and that the vertical component Bh is equal to Ae. By examining the figure carefully we find the following strains on the different parts:

Strains of Extension

on AE	is	Ab
on EG	is	Ec
on FG	is	$Fu = Ab$
on GI	is	$2Ec$
on HI	is	$Fu = Ab$
on IB	is	$3Ec$
on KB	is	$Bg = Fu = Ab$
on AF	is	Aa
on FH	is	$Aa - Ft$
on HK	is	$Aa - 2Ft$
on point K	is	$Aa - 3Ft = 0$
Hor. strain on point B	is	$4Ec = Bw = ef$
Vert. " " "	is	$Bh = Ac$
Total " " "	is	Af
on AC	is	$An = Cp$
on point C	is	$An = Cp$

Strains of Compression

on EF	is	$Fv = Ae$
on GH	is	$Fv = Ae$
on IK	is	$Fv = Ae$

We observe that the strains in the segments of the Upper Chord accumulate towards B, while those in the segments of the Lower Chord decrease toward the point K, on which there is no force acting. In practical construction, however, there would be a shearing strain on the pin, or other similar part used at this point to connect together the post, chord, and tie. This is one of those cases in which a force theoretically neutralized must nevertheless be considered by the bridge-builder, on account of the necessity of using pins or other joining members to connect together the main members through which come the strains that neutralize each other. It is a mechanical necessity whose value can be accurately determined.

With the combination shown in Figure 8, we see that the due proportions of the weight have been carried to the abutments; but that, besides having to sustain the vertical pressure of these weights, the abutments have to resist horizontal forces tending to overturn them. The discussion has shown us that whatever the amount of the weight, and whatsoever its location on the truss, the horizontal forces developed are equal and contrary. This at once suggests the idea of making these equal and opposite forces neutralize each other, thereby relieving the supports from all horizontal strain.

If, therefore, we cast loose from B and C, as far as resistance to horizontal strains is concerned, and insert a strut between E and C, it is evident that the horizontal forces in the Top Chord will accumulate towards E, the new point of resistance, instead of towards B, which is now free; and the opposite horizontal forces will neutralize each other through EC. We thus obtain

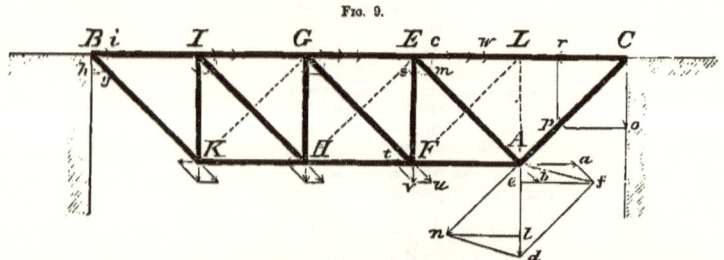

Fig. 9.

We readily see that the only change caused in the character of the strains developed is to make all the strains in the segments of the Top Chord strains of compression, instead of strains of extension, and to cause the strains to accumulate towards L instead of towards B. Neither the amount nor the character of the other strains is affected by the change.

The new strains on the Top Chord are:

on BI a strain of compression equal to Bi.
on IG "   "   " $2Bi$.
on GE "   "   " $3Bi$.
on EC "   "   " $4Bi = Cr = nl = ef$.

If, now, the intermediate parts KG, HE, FL, and AL be inserted, it is manifest that, whatever may be the initial strains, the weight is resolved into two components, An and Af. As the truss should be symmetrical, we will make the new inclined parts ties, and the vertical part a post. We find the first decomposition at A necessarily unchanged. At E, Em will naturally be resolved into Ec and Es, components in the direction of the nearest members that can support the strain. The ties will not sustain a compressive strain; and if such a strain should, from any defect of a joint, come on a tie, it will necessarily yield until the post comes into play. A brace cannot be compressed from one end and extended from the other, as the coexistence of the two cases requires the fixedness of the two ends, and, at the same time, their freedom; which is impossible.

Exactly how it is that the weight at once resolves itself into the proper proportions for transmission to the points of support, sending to each abutment exactly the amount, measured horizontally and vertically, that should go there, is not definitely known. That such is ultimately the case, both theory and experiment prove; but what preliminary and almost instantaneous adjustments are undergone before the division is effected, is not clear; nor, indeed, is it essential that we should know.

The following Figures show the changes in the resolutions of a given weight caused by varying the distance to one of the abutments:

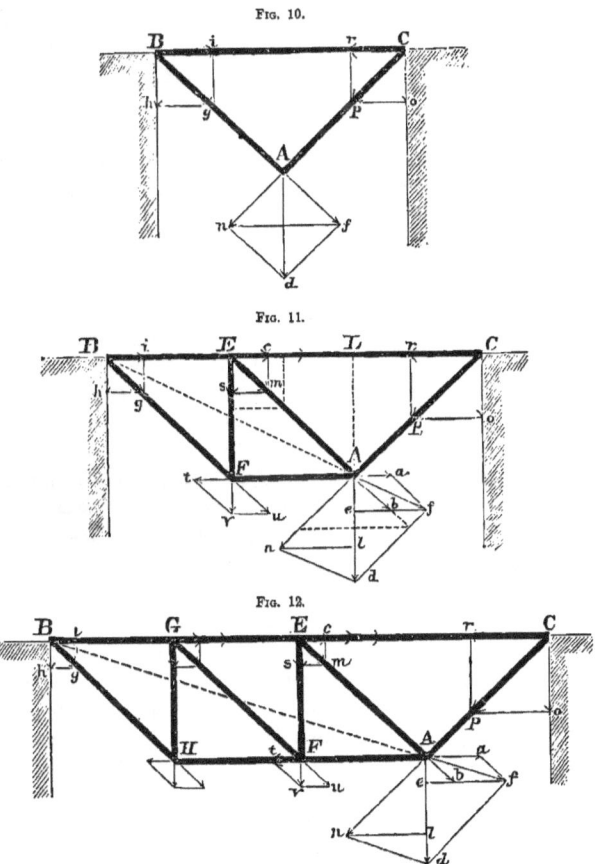

Fig. 10.

Fig. 11.

Fig. 12.

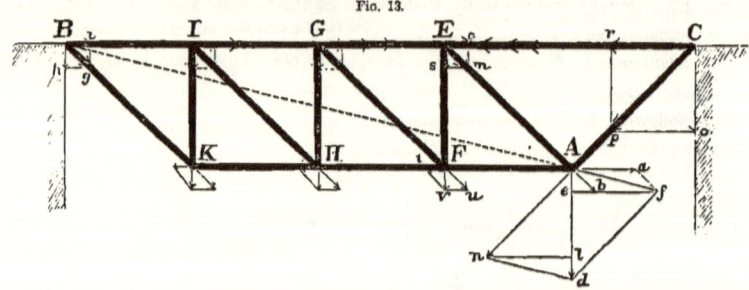

Fig. 13.

Figure 10 needs no explanation.

In discussing Figure 11, let us see if we cannot corroborate our previous conclusions, and prove, by a different course of reasoning, that the weight at A must be resolved along the tie AC and the theoretical line AB.

Suppose that the weight is at once resolved into components along the prolongations of AC and AE, as indicated by the dotted lines. The parallelogram of forces at A shows at once that these components are equal in themselves, and in their horizontal and vertical components. Neither component develops any tension at A, as they are transmitted at once to C and E. At E, horizontal and vertical components are developed equal to those at C. But E being an unsupported point, the vertical component develops horizontal and vertical components at F, and the vertical component at F develops horizontal and vertical components at B equal to those first obtained at E. We thus have on EC a compression from left to right equal to double the compression from right to left. The excess of compression from the left must be sustained by the abutment C, or motion will take place. Moreover, we find B sustaining the same vertical weight as C, and we find at F a horizontal force acting to the left that is resisted by nothing, and that will also cause motion. Both theory and practice show that our conclusions are absurd, and that we have evidently incorrectly resolved the original weight. We know that B should sustain only one-third the original weight, and therefore that the vertical component of the portion of the weight transmitted to B should be one-half the vertical component of the weight transmitted to C. Also, the portion of the weight that goes to B must develop a tension at A to neutralize the tension that will necessarily be developed at F. And, in addition, in order to prevent motion in the Top Chord, the force that goes up AE must be of such magnitude that its horizontal component shall be one-half of the horizontal component of the force that goes up AC as the former is doubled by the compression from B. None of these conditions being fulfilled by the resolution just made, we therefore see that it must be erroneous.

From the necessary conditions just indicated we can now determine the proper line of resolution for the component to the left. Laying off $A\,e$ equal to one-third of the weight $A\,d$, and $A\,l$ equal to two-thirds of $A\,d$, we evidently have the vertical components of the two parts of the original weight, the smaller being the portion that is to go to B. Drawing the horizontal line $l\,n$ through l, until it meets $C\,A$ produced, we get $A\,n$, the true strain that goes to C. Through e draw the horizontal line $e\,f$, making it equal to $l\,n$, and produce $E\,A$ until it meets $e\,f$. In the similar triangles $A\,l\,n$ and $A\,e\,b$, $A\,e$ is one-half of $A\,l$, and therefore $e\,b$ is one-half of $l\,n$, and $e\,f$ is bisected at b. Completing the parallelogram $A\,a\,f\,b$ we get $A\,a = b\,f = e\,b$. The line $A\,f$ fulfils all the conditions which we have seen, à priori, to be necessary as the line of resolution for the component that goes to the left. The sum of the two compressions which it gives in the Top Chord is $E c + B i = 2\,e\,b = e\,f = l\,n$; the tension it develops at F is $F t = A a$, the two neutralizing each other; and $A\,e$, the portion of the weight which it transmits to B, is one-half of $A\,l$, as it should be. This line $A f$ passes through B, because $A\,L\,E$ and $A\,e\,b$ being similar triangles with parallel sides, $A\,L\,B$ and $A\,e f$ must be so also; $A f$ is therefore parallel to $A\,B$, and as they meet in A they must be one and the same line.

Comparing together Figures 10, 11, 12, and 13, we obtain the following conclusions for symmetrical trusses of this character, sustaining a weight not at the middle point.

For the same weight the further we remove one of the points of support, leaving the other unchanged, the greater the portion of the weight borne by the unmoved abutment, the greater the strain on the Tie connecting it with the weight, and the greater the horizontal force of compression on the Top Chord at the abutment end of this tie.

Also, on the other side, we see that each Post is compressed by a force exactly equal to the portion of the weight originally to be transmitted, and that each Tie has the same strain of extension, and develops the same force of compression on its own segment of the Top Chord. These compressions are equal to each other, and their sum is equal to the entire horizontal component that originally required transmission to the left, and it neutralizes the horizontal force developed on the right. On the Bottom Chord we see that the greatest tension is equal to either horizontal component of the weight diminished by one of the top compressions ($e b = e m$), and that the tensions diminish toward the further abutment, and at the foot of the last Tie become 0, thereby showing that all the horizontal forces developed in the process of transmitting the portions of the weight to each abutment, neutralize each other through the horizontal members of the truss, leaving no unneutralized force to cause motion. We also see that the weights are transmitted only by the vertical or inclined parts, the horizontal parts being only used for the necessary decompositions required to pass the weights from inclined to vertical members, and vice versa.

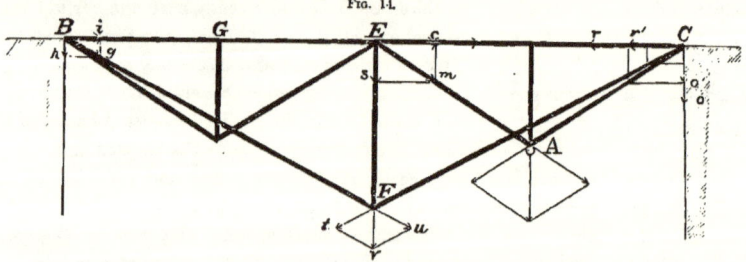

Fig. 14.

There is a class of so-called trusses, distinct from all others, that are more properly trussed girders. Figure 14 represents one of the class which we will have to discuss in detail further on. It is manifest that, there being but two members meeting at A, the component parts of the weight must pass up those two. As they make equal angles with each other, the weight is halved. So far C has received but half the weight, when the relative distances from A to B and C show that it should receive three-fourths. The component at E is resolved vertically and horizontally. The vertical component transmitted to F is resolved into Ft, that goes to increase the forces at C; and into Fu, that goes to B. An examination of the figure will show that, as before, the horizontal forces developed in the Top Chord are equal and opposite, while the vertical forces at the abutments are inversely proportional to the distances of the weight from those abutments. In this combination there is no Bottom Chord, and the principle of the counter-balancing of equal weights is inapplicable. The necessity of equalizing the horizontal forces in the Lower Chord, which appeared in the other combinations, does not hold in this truss, and hence the resolution is much simpler.

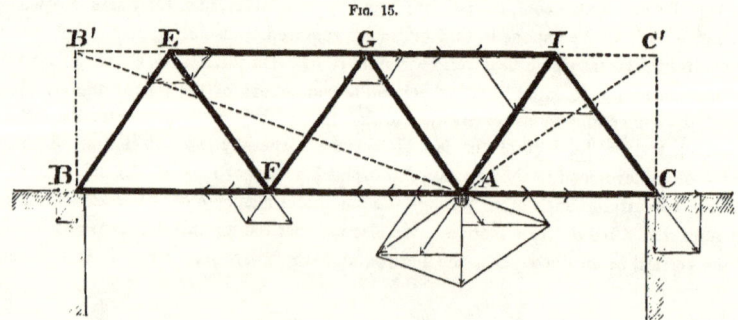

Fig. 15.

In Figure 15 the points B' and C', through which we draw the lines from A, along which the weight is to be resolved, have no physical existence. This, however, does not affect the principle of the resolution, for it takes place along $A B'$ and $A C'$ as if the Top Chord extended to B' and C'. We might at first imagine that the resolution would take place along $A I$ and $A E$, but we know beforehand that one third of the weight goes to B, and two-thirds to C, which division is accomplished by the resolution along $A B'$ and $A C'$, while the resolution along $A E$ and $A I$ would send one-fourth of the weight to B and three-fourths to C—a manifest mistake.

Extra Strains.

In iron bridges of any length it is impracticable to cast the entire Top Chord in one piece; it must be made up of segments, and a convenient and natural mode of division is to make each segment the length of a panel. The Ties, Posts, and segments of the Top Chord for each panel meet in a common point. It is of the greatest importance to keep the points of contact of the segments of the Top Chord in the same horizontal line. If the lines connecting the points of contact, which are the lines of direction of the forces acting on the segments of the Top, are not portions of the same right line, they must develop upward or downward components at their points of meeting.

These points of meeting of the segments ought to be identical with the centres of their ends, but from various causes they may be eccentric. Imperfections in workmanship, the ends not being truly faced perpendicular to the axes, may cause the centre of pressure to be above or below the centre of figure of the end, or this eccentricity may arise from defective adjustment of the bridge, causing sagging in some places and a tendency to rise in others.

In most bridges these possible extra strains will be found to give but a small increase in some of the maxima strains as calculated without them, and this discussion might safely be omitted were it not for the Fink and Bollman bridges, especially the latter. Experience has shown that in it certain Posts and Panel Ties are absolutely required upon which no direct strains (excepting the weight of the Top segment) can be found. The strains that necessitate the Panel Ties of this bridge, and that fix the dimensions of both Posts and Ties, must be the extra strains which have been indicated. In order, therefore, to make a fair comparison between the seven bridges selected we have been compelled to allow for these additional strains in them all.

In most trusses each segment of the Top has a different strain from its neighbor. The difference arises from the fact that in each panel additional strains are brought up by the Ties. Whenever we discuss such a truss we will always choose for the strain at any joint the lesser of the strains on the segments that form this joint, as that is the force with which the segment with the greater strain is pushed, and hence any tendency to divergence must be referred to this force that acts to cause it. The ends of the Top Chord being free, and the line connecting them being *the line of the Chord*, it is manifest that no extra strains are to be provided for at these points.

The effect of a divergence from the line of the Chord will be to develop an upward or downward tendency at any joint equal to the force of compression in the segment furthest from the centre of compression multiplied by the tangent of the angle between the axis of the segment and the line of direction of the compression acting through it. The greatest value that can be given to this tangent is equal to the radius of the end of the segment.

In allowing for this extra strain in the bridges which we intend examining, we will have to assume that the external diameters of all Top Chords are equal, a difference in strain being met by a difference in thickness, and that the panel lengths are the same in all. This uniform radius we will take at 9" for a 200-foot span, the panel length being 12' 6". Then in the triangle represented in Fig. 16, we find that the third side is equal to 150".3. But the hypothenuse represents the compression on the segment of the Top. The downward tendency in this case would evidently be equal to 9 divided by 150.3, or $\frac{1}{17} \times$ the compression on the segment.

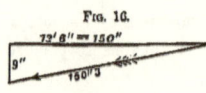
Fig. 16.

Whenever, therefore, this extra strain that *may* arise would increase the total amount of strain we must add it to the maximum direct strain, and provide for it as for the regular strain.

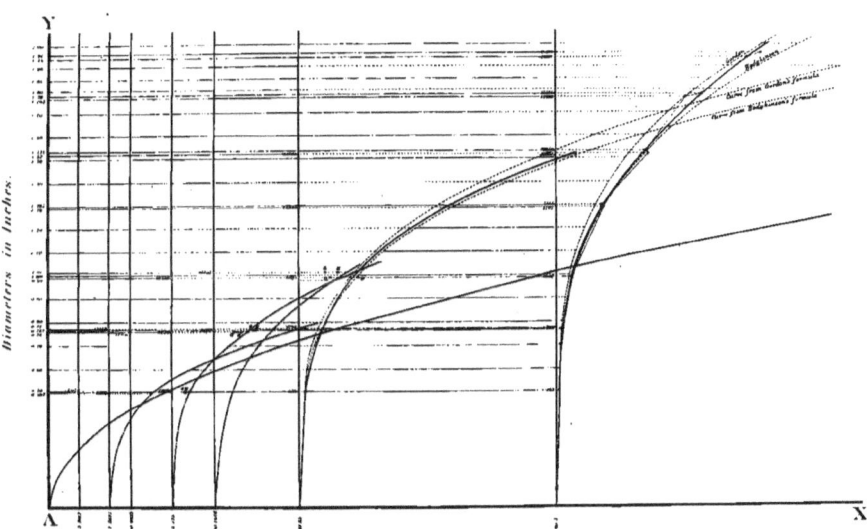

Hodgkinson's Experiments on Cast Iron Pillars with Rounded Ends.

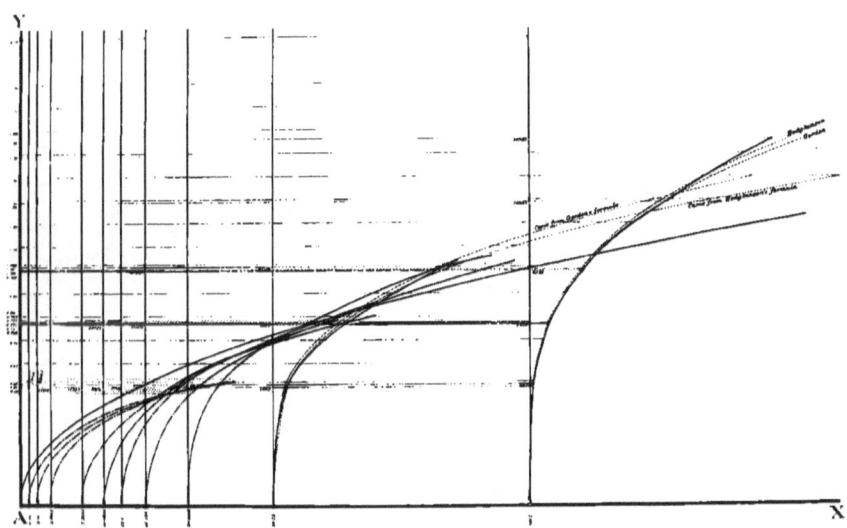

HODGKINSON'S EXPERIMENTS ON CAST-IRON PILLARS.

In order to make estimates of the quantity of iron needed in the bridges which we propose to examine, it is necessary to adopt some formulas for the strength of cast and of wrought iron. As the latter is only used for tension in the bridges which we have selected, and as its strength to resist tension is universally acknowledged to vary as its minimum cross-section, the formula for wrought-iron becomes simple enough. With cast-iron it is different.

After much theoretical investigation, Mr. Eaton Hodgkinson determined to ascertain from experiment the actual breaking weight of pillars and beams of cast-iron, varying the conditions so as to determine *the law of the strength*, if possible. We are only concerned, however, with his experiments on the resistances of pillars. The Posts and Struts of an iron bridge, and the segments of the Top Chord, are subjected to the same strains as Hodgkinson's pillars, and, therefore, our calculations must be based on the results of those experiments.

In seeking for the law of the strength of pillars, Hodgkinson confined himself to one variety of iron, assuming that the strength of any other variety might be determined at once by suitably modifying a constant coefficient. He chose for his experiments the iron known as "Low Moor No. 3," on account of its general uniformity of strength. He experimented on solid and on hollow pillars, concluding that the strength of a hollow pillar was equal to the difference between the strengths of two solid pillars of the given length, having for their diameters the external and the internal diameters of the given hollow pillar. It is proper to add, however, that he uses a smaller numerical coefficient for the hollow than for the solid pillar.

We will first examine the experiments on solid pillars, though in applying the formulas to actual bridge calculation we will use those for hollow pillars, as the saving of metal in hollow over solid pillars of equal length has caused them to be generally adopted in bridges.

Mr. Hodgkinson tested two classes of solid pillars, one with ends hinged or rounded, and the other with ends flat. For the Posts of bridges we must obviously use the formula for round-end pillars, and the same formula seems to be the most convenient one for the segments of the Top. It is impossible to reproduce in the Top Chord of a bridge the conditions of perfect immobility, and certainty of the forces acting through the axis, that existed in the flat-end pillars on which Hodgkinson experimented. A reference to the report of his experiments, with a glance at the drawings of the testing machines, shows this conclusively. But he found that when a flat-end pillar is subjected to slight movements, and when the line of direction of the strain does not coincide exactly with its axis, its strength is about one-third of what it is when the ends are fast. But the strength of round-end pillars is also about one-third that of immovable flat-end ones. Hence we may either calculate the strength of the top segments from the flat-end formula, and divide by three, or we may use the round-end formula. For our present purpose it is simpler to use but one formula for cast-iron, and, therefore, we use the formula for round-ended pillars.

In reference to the shape of the formula itself something should be said. The number of experiments on long pillars with round ends (taking the mean results only for the same length and diameter) was but fifteen—a number disproportionately small on which to base the innumerable cases of the use of cast-iron pillars in bridge-building, architecture, etc.; and yet these are the only reliable published data that we have. From them Hodgkinson and Gordon deduced the formulas for the breaking weights of pillars that are in common use.

Hodgkinson's formula for the breaking weight of a solid pillar *with flat ends*, whose length is not less than 30 times its diameter, is

$$W = 98922 \frac{d^{3.55}}{l^{1.7}}.$$

For a pillar *with rounded ends*, whose length is not less than 15 times its diameter, he has

$$W = 33379 \frac{d^{3.76}}{l^{1.7}};$$

in both of which

$W =$ the breaking weight in pounds;
$d =$ the diameter in *inches*;
$l =$ the length in *feet*.

For a short pillar, whose length in diameters is less than 30 and 15 respectively, he has the supplementary formula,

$$y = \frac{bc}{b + \frac{3}{4}c};$$

in which

$y =$ the actual breaking weight in pounds;
$b =$ the breaking weight in pounds, as given by one of the preceding formulas;
$c =$ the crushing weight per cross-section, assumed at 109,801 lbs. per square inch.

As we shall have frequent occasion to use this supplementary formula, especially in getting the breaking weights of top segments, which often come under the class of "short" pillars, it will be convenient to combine it at once with the flat and round end formulas for "long" pillars, and thus obtain but a single formula for "short" pillars of either class.

In the supplementary formula
$$b = W$$
$$c = 109801 \times \text{cross-section}.$$
Hence, for *flat ends*,
$$b = 98922 \frac{d^{2.55}}{l^{1.7}}$$
$$c = 109801 \times \tfrac{1}{4}\pi d^2$$
And we get in the formula
$$y = \frac{98922 \times \frac{d^{2.55}}{l^{1.7}} \times 109801 \times \tfrac{1}{4}\pi d^2}{98922 \times \frac{d^{2.55}}{l^{1.7}} + \tfrac{3}{4} \times 109801 \times \tfrac{1}{4}\pi d^2}$$

$$y = \frac{\tfrac{1}{4}\pi \times 109801 \times \frac{d^{2.55}}{l^{1.7}}}{\frac{d^{2.55}}{l^{1.7}} + \frac{3\pi}{16} \times \frac{109801}{98922}}$$

$$y = \frac{\tfrac{1}{4}\pi \times 109801 \times d^2}{1 + \frac{3\pi}{16} \times \frac{109801}{98922} \times \frac{l^{1.7}}{d^{1.55}}} = \frac{86237.5\, d^2}{1 + 0.65383 \times \frac{l^{1.7}}{d^{1.55}}}$$

Similarly, for short *round-end* pillars,
$$y = \frac{\tfrac{1}{4}\pi \times 109801 \times d^2}{1 + \frac{3\pi}{16} \times \frac{109801}{33379} \times \frac{l^{1.7}}{d^{1.76}}} = \frac{86237.5\, d^2}{1 + 1.93769 \frac{l^{1.7}}{d^{1.76}}}$$

In using these two formulas, we will substitute W for y to make them conform better with the others.

For HOLLOW *flat-end* pillars Hodgkinson gives
$$W = 99318 \frac{D^{2.55} - d^{2.55}}{l^{1.7}}$$
And for HOLLOW *round-end* pillars
$$W = 29074 \frac{D^{2.76} - d^{2.76}}{l^{1.7}}$$

In each case, excepting only a slight change in the value of the numerical coefficient, he makes the strength of a hollow pillar equal to the difference between the strengths of two solid pillars of the same length, and whose diameters are respectively equal to the external and internal diameters of the hollow pillar.

Making the necessary modifications due to the changed coefficients, in the formulas for short lengths, we get

For *flat ends*, less than 30 diameters long,
$$W = 86237.5 \left(\frac{D^2}{1 + 0.651223 \frac{l^{1.7}}{D^{1.55}}} - \frac{d^2}{1 + 0.651223 \frac{l^{1.7}}{d^{1.55}}} \right)$$

And for *round ends*, less than 15 diameters long,
$$W = 86237.5 \left(\frac{D^2}{1 + 2.2246 \frac{l^{1.7}}{D^{1.76}}} - \frac{d^2}{1 + 2.2246 \frac{l^{1.7}}{d^{1.76}}} \right)$$

These formulas are collected together elsewhere.

Gordon's formula (slightly transformed) for a solid cylindrical pillar *with flat ends* is

$$W = \frac{8000000 \pi d^2}{400 + \frac{l^2}{d^2}} = \frac{25132741 \, d^2}{400 + \frac{l^2}{d^2}}$$

And for one *with rounded ends* is

$$W = \frac{2000000 \pi d^2}{100 + \frac{l^2}{d^2}} = \frac{6283185 \, d^2}{100 + \frac{l^2}{d^2}}$$

In which $W =$ the breaking weight in pounds.

$d =$ the diameter in *inches*.

$l =$ the length in *inches*.

Gordon has but one formula for each kind of solid pillar, and his lengths and diameters are referred to the same unit. Hodgkinson takes his lengths in feet and his diameters in inches, and uses a supplementary formula for short pillars.

Gordon's formula appears to be an adaptation for engineers who are not familiar with the use of logarithms. This makes it inconvenient for use in the investigations necessary for the discussions that follow, and Hodgkinson's formula has therefore received the preference. Moreover, the formula of the latter is in form more in agreement with what theory would indicate *à priori* as the probable shape of the formula—the breaking weights varying directly as some power of the diameter, and inversely as some power of the length. For round-end pillars (with which we are particularly concerned) it gives results nearer those found by actual trial than Gordon's formula. This can readily be seen in the tables that follow, in which the breaking weights determined by experiment have been compared with those given by the formulas, and also by examining the curves constructed in the plates. Neither formula exactly reproduces the experiments, Gordon's always giving for round-end pillars less than the true breaking weight. The difficulty of determining an accurate formula comes from the fact that *the experiments are too few in number to determine a law*, merely sufficing to indicate the existence of one.

On the accompanying diagrams all of the experiments on both long and short pillars are recorded. Assuming three axes of co-ordinates, and laying off the lengths on the axis of X, the diameters on the axis of Y, and the breaking weights on the verticals at their intersections, we obtain by connecting the tops of these verticals a surface whose equation will always give us z, the breaking weight, for any values of x and y. This surface meets the plane of reference in the axis X.

We know that when the lengths are very small the pillar is destroyed by crushing without flexure, and that the weights are proportional to the cross-sections, or to the squares of the diameters. Hence the curve cut out of the surface by the plane YZ has an equation of the form

$$W = A \times \tfrac{1}{4}\pi y^2.$$

The experiments which determined this form of equation, and the value of the constant A, were all made on small pillars with flat ends. Our experiments are not sufficiently numerous to enable us to decide positively; but we may fairly assume that for very short pillars the breaking weights are the same whether the ends be flat or round. In comparing the experiments we find that for "long" pillars of the same length and diameter the flat-ended will sustain thrice the load of the round-ended. As we diminish the lengths the ratio between the strengths of the two kinds of pillars becomes less than 3, then 2, then less than 2, and our last experiments on a pillar 0.5 inch in diameter 3.78 inches long gave 17468 as the breaking weight of the flat-ended, and 15107 as that of the round-ended. We may therefore conclude that for still smaller lengths the shape of the ends does not affect the breaking weight.

Writing z for W, and giving A its experimental value, we have

$$z = 109801 \times \tfrac{1}{4}\pi y^2 = 86237.5\, y^2$$

$$y^2 = \frac{1}{86237.5}\, z$$

This is the equation of a parabola passing through A, the origin of co-ordinates, of which the axis Z is the axis, and $\frac{1}{86237.5}$ is the parameter of the axis.

This parabola could be described from its equation, but as we have different scales for diameters and breaking weights, the equation would require transformation before the parabola could be constructed graphically. It will be sufficiently accurate if we construct it by points. Solving the equation for different values of y we obtain

y	z	y	z	y	z	y	z
0.025	54	0.3	7761	0.65	36435	1.29	143508
0.05	216	0.35	10564	0.7	42256	1.295	144622
0.075	485	0.4	13798	0.76	49811	1.52	199243
0.1	862	0.45	17463	0.767	50733	1.535	203195
0.125	1347	0.497	21301	0.77	51130	1.765	268049
0.15	1940	0.5	21559	0.99	84521	1.78	273235
0.2	3450	0.55	26087	1.01	87971	1.94	324563
0.25	5390	0.6	31046	1.2	124182	1.96	331290

By which we plat the curve.

Our data seem insufficient to determine with any accuracy more than one cross-section of the surface—that made by a plane parallel to YZ, and at a distance from the origin of 60.5 inches. Constructing the points of this section from the actual breaking weights, and connecting them by right lines, we find that they roughly outline a curve. Drawing this curve by the eye, we obtain one of the cross-sections of the surface whose equation we are seeking. In attempting to calculate the equation of this curve, we must take the ordinates given by the drawing, some of which are necessarily a little different from the values as given by the experiments. Bearing in mind the many extraneous causes that may operate to injure a casting, but knowing that there must be *some* law for the strength of cast-iron pillars, we may safely assume that our corrected breaking weights for the round-end pillars are as nearly right as we can make them with our present limited supply of facts. The most correct way to get the equation of this curve is probably by the formulas for interpolation. The work, however, is very long and tedious, and arithmetical errors can scarcely be avoided. We have twice gone through with the whole calculation, and each time have obtained an unsatisfactory result. The experiments on flat-end pillars of 60½ inches in length are so few as to prevent much discussion.

Curves cut out by planes parallel to the plane XZ have their maxima ordinates at their points of intersection with the curve cut out by the plane YZ, the ordinates decreasing as the curve recedes from YZ, until for great values of l the ordinates become 0, and the curve attains the horizontal plane XY—or, in other words, the pillar breaks under its own weight. In most pillars the weight of the pillar itself need not be taken into account, and W represents the extraneous weight that crushes it. When the pillar is so long that its own weight is great in proportion to its breaking weight, we must consider W as composed of the weight of the pillar itself above the expected point of fracture, and of the outside weight laid on it. If then we calculate W from the formula, and subtract a part of the weight of the pillar, depending upon the position of the point which we select as that of fracture, we will get the weight that laid on the pillar at its upper end, will just break it. It is evident that for some lengths the weight of the pillar will be so great that when the proper portion is subtracted from W there will be a remainder equal to or less than 0; the pillar then will not stand under its own weight.

The horizontal contour lines of the surface can be determined by finding the points of equal breaking weights on the curves of section of each length, and connecting them. We can see that they are curved lines, close together as they start from the plane YZ, and gradually diverging as they recede from this plane. As x (or l) increases, the tangents to the contour lines become more and more nearly parallel to the axis of X.

If enough additional experiments were made to give a fair number of cross-sections of the surface, so that we could draw the contour lines with some degree of accuracy, it would not be absolutely necessary to determine the equation of the surface, as the breaking weight for every value of x and y could be readily determined by simple inspection. Some experiments should be made on pillars more near in size to those actually used than were those on which Hodgkinson experimented. The length of his longest pillar was only 5 feet.

By comparing the results for round-end pillars 60.5 inches long as obtained by experiment, and by Hodgkinson's formula, we find that the greater the diameter, and therefore the nearer that the ratio between it and the length approaches to 30, the more the results vary; those given by the formula being in excess of the actual breaking weights. Either the co-efficient is too great or the power of the diameter. A similar state of affairs holds with regard to the experimental curve for lengths of thirty and a quarter inches. The curves from the formula and the experimental curves agree very well for the smaller values of d, but in each case the curve from the formula passes above that from experiment as the value of d increases. Gordon's formula gives a curve that passes below in both cases. It seems quite probable that a further investigation would give some formula that would more nearly reproduce the experiments than the two which we have; but the matter is rather incidental to the present subject, having been only undertaken to show the value of our present formulas for the strength of cast-iron. To determine a new and more exact formula may be the subject of a subsequent investigation, but it cannot be commenced now.

The exponent of d can hardly be constant, for we have just shown that when $l = 0$ d equals 2, and when $l = 60.5$ $d = 3.76$, or thereabouts, for round-end pillars, and 3.55 for flat-end ones. In both cases the change from 2 to 3.76, or 3.55, must be gradual, as there can be no abrupt break in the surface upon which are found all the values of W. It is hardly probable that the value of d for $l = 60.5$ would be its maximum. Moreover, the fact that Hodgkinson's formula gives greater values than experiment does for the breaking weight when d is large, l being 60½ inches, would seem to show that the exponent of d should decrease for a given length of pillar, as d itself increases. Or this decrease may come from the coefficient which may be a decreasing function of d. These, however, are only suggestions to assist future inquiry.

The entire discussion shows that there is a great insufficiency of reliable experiments on this point, and also that while much remains to be done to perfect our knowledge, it is yet possible with the means at hand to obtain more accurate formulas than either Hodgkinson's or Gordon's. Nevertheless, while awaiting the future formula, we will adopt those of Hodgkinson as having the greatest amount of public confidence.

The breaking weights as obtained from experiment, and by calculation, for both kinds of pillars, are tabulated in the following:

Solid Pillars with Flat Ends.

Long Pillars.

Length in Diamet's.	Length in Inches.	Diameter in Inches.	Experimental Breaking Weights.	Breaking Weights as given by Hodgkinson's Formula.	Difference. From Experimental Breaking Weight.	In per Cent.	Breaking Weights as given by Gordon's Formula.	Difference. From Experimental Breaking Weight.	In per Cent.
33	60.5	1.80	....	50949	....	....	53232	....	....
36	60.5	1.70	....	41592	....	....	43584	....	....
38	60.5	1.60	....	33538	....	....	35162	....	....
39	60.5	1.56	28962	30656	+1694	+6	32122	+3160	+11
40	60.5	1.50	....	26671	....	....	27901	....	....
43	60.5	1.40	....	20877	....	....	21725	....	....
47	60.5	1.29	16064	15614	−450	−3	16089	+25	+¼
50	60.5	1.20	....	12078	....	....	12302	....	....
55	60.5	1.10	....	8869	....	....	8879	....	....
61	60.5	1.00	....	6323	....	....	6190	....	....
61	60.5	0.997	6238	6256	+18	+⅓	6120	−118	−2
67	60.5	0.90	....	4350	....	....	4139	....	....
78	60.5	0.80	....	2863	....	....	2629	....	....
79	60.5	0.77	2456	2500	+44	+2	2267	−189	−8
86	60.5	0.70	....	1782	....	....	1565	....	....
101	60.5	0.60	....	1031	....	....	856	....	....
119	60.5	0.51	487	579	+92	+19	452	−35	−7
30	30.25	1.01	20310	21282	+972	+5	19767	−543	−3
30	30.25	1.00	....	20543	....	....	19112	....	....
34	30.25	0.90	....	14133	....	....	13308	....	....
38	30.25	0.80	....	9303	....	....	8791	....	....
39	30.25	0.77	8811	8123	−688	−8	7668	−1143	−13
43	30.25	0.70	....	5791	....	....	5431	....	....
50	30.25	0.60	....	3350	....	....	3076	....	....
61	30.25	0.50	1662	1754	+92	+6	1548	−114	−7
40	20.1666	0.51	3830	3749	−81	−2	3329	−501	−13
30	15.125	0.51	6764	6114	−650	−10	5109	−1655	−25

IRON TRUSS BRIDGES FOR RAILROADS. 33

SOLID PILLARS WITH FLAT ENDS—*Continued.*

Short Pillars.

Length in Diamet's.	Length in Inches.	Diameter in Inches.	Experimental Breaking Weight.	Breaking Weight as given by Hodgkinson's Formula.	DIFFERENCE.		Breaking Weight as given by Gordon's Formula.	DIFFERENCE.	
					From Experimental Breaking Weight.	In per Cent.		From Experimental Breaking Weight.	In per Cent.
20	30.25	1.50	....	72418	....	....	70099	....	....
22	30.25	1.40	....	58916	....	....	56825	....	....
23	30.25	1.30	....	47068	....	....	45116	....	....
25	30.25	1.20	....	36813	....	....	34952	....	....
28	30.25	1.10	....	28080	....	....	26301	....	....
20	20.1666	1.022	31804	*35631	+3827	+12	33255	+1451	+5
26	20.1666	0.777	15581	*15604	+23	+⅒	14133	−1448	−9
15	15.125	1.00	40250	*43797	+3547	+9	43205	+2955	+7
20	15.125	0.775	21509	*21241	−268	−1	19331	−2178	−10
15	12.1	0.785	24287	*27043	+2756	+11	24291	+4	+$\tfrac{1}{1000}$
24	12.1	0.50	7195	*7328	+133	+2	6375	−820	−11
13	10.0833	0.768	25923	*29363	+3440	+13	25899	−24	−$\tfrac{1}{1000}$
20	10.0833	0.50	8931	*8872	−59	−1	7789	−1142	−13
10	7.5625	0.777	32007	*36130	+4123	+13	30737	−1270	−4
15	7.5625	0.50	11255	*11508	+253	+2	9993	−1262	−11
8	3.7812	0.50	17468	*16992	−476	−3	13743	−3725	−21
4	2.0	0.52	22867	21479	−1388	−6	16384	−6483	−29
2	1.0	0.52	24616	22720	−1896	−8	16834	−7782	−33

* These are Hodgkinson's calculations as given in his Report.

4

Solid Pillars with Rounded Ends.

Long Pillars.

Length in Diamet's.	Length in Inches.	Diameter in Inches.	Experimental Breaking Weight.	Corrected Breaking Weight.*	Breaking Weights as given by Hodgkinson's Formula.	Difference.		Breaking Weight as given by Gordon's Formula.	Difference.	
						From Experimental Breaking Weight.	In per Cent.		From Experimental Breaking Weight.	In per Cent.
30	60.5	2.00	....	....	28905	....	....	24760	....	....
31	60.5	1.96	24291	23580	26790	+3210	+14	22927	−653	−3
31	60.5	1.94	22811	....	25777	+2966	+13	22053	−758	−3
33	60.5	1.85	....	....	21561	....	....	18388	....	....
34	60.5	1.78	17564	....	18650	+1086	+6	15856	−1708	−10
34	60.5	1.765	15560	17130	18066	+936	+5	15352	−1778	−10
36	60.5	1.70	....	....	15689	....	....	13288	....	....
38	60.5	1.60	....	....	12491	....	....	10515	....	....
39	60.5	1.535	10650	....	10687	+37	+⅓	8954	−1696	−16
40	60.5	1.52	10861	10320	10300	−20	−⅓	8618	−1501	−14
40	60.5	1.50	....	....	9799	....	....	8187	....	....
43	60.5	1.40	....	....	7560	....	....	6259	....	....
47	60.5	1.295	5465	5660	5639	−21	−⅓	4616	−1044	−18
47	60.5	1.29	5707	5570	5558	−12	−⅓	4547	−1023	−18
50	60.5	1.20	....	....	4373	....	....	3425	....	....
55	60.5	1.10	....	....	3053	....	....	2433	....	....
61	60.5	1.00	....	....	2134	....	....	1671	....	....
61	60.5	0.99	1902	....	2054	+152	+8	1606	−296	−16
67	60.5	0.90	....	....	1436	....	....	1102	....	....
78	60.5	0.80	....	....	922	....	....	691	....	....
79	60.5	0.77	780	....	799	+19	+2	594	−186	−24
121	60.5	0.50	143	....	158	+15	+10	107	−36	−25
15	30.25	2.00	....	....	93912	....	....	76446	....	....
16	30.25	1.90	....	....	77439	....	....	64169	....	....
17	30.25	1.80	....	....	63194	....	....	53232	....	....
18	30.25	1.70	....	....	50973	....	....	43584	....	....

* These "corrected" breaking weights are the ordinates of the curve of the experimental breaking weights as constructed on the diagram.

IRON TRUSS BRIDGES FOR RAILROADS.

SOLID PILLARS WITH ROUNDED ENDS—*Continued.*

Long Pillars—*Continued.*

Length in Diamet's.	Length in Inches.	Diameter in Inches.	Experimental Breaking Weight.	Corrected Breaking Weight.	Breaking Weights as given by Hodgkinson's Formula.	DIFFERENCE.		Breaking Weight as given by Gordon's Formula.	DIFFERENCE.	
						From Experimental Breaking Weight.	In per Cent.		From Experimental Breaking Weight.	In per Cent.
19	30.25	1.60	....	....	40583	....	....	35163	....	....
20	30.25	1.52	32531	....	33464	+933	+3	29264	−3267	−10
22	30.25	1.40	....	....	24563	....	....	21725	....	....
23	30.25	1.29	17235	....	18058	+823	+5	16089	−1146	−7
25	30.25	1.20	....	....	14209	....	....	12302	....	....
28	30.25	1.10	....	....	9919	....	....	8879	....	....
30	30.25	1.00	....	....	6932	....	....	6190	....	....
30	30.25	0.99	6105	....	6675	+570	+9	5958	−147	−2
34	30.25	0.90	....	....	4664	....	....	4139	....	....
38	30.25	0.80	....	....	2906	....	....	2629	....	....
39	30.25	0.77	2726	....	2595	−131	−5	2267	−459	−17
43	30.25	0.70	....	....	1813	....	....	1565	....	....
50	30.25	0.60	....	....	1016	....	....	856	....	....
61	30.25	0.50	539	....	512	−27	−5	418	−121	−23
20	20.1666	1.01	15737	....	14337	−1400	−9	12853	−2884	−19
26	20.1666	0.767	6602	....	5093	−1509	−23	4671	−1931	−29
15	15.125	0.99	19752	....	21686	+1934	+10	18470	−1282	−6
20	15.125	0.76	9223	....	8025	−1198	−13	7316	−1907	−21
30	15.125	0.5	1904	....	1663	−241	−13	1547	−357	−19

Short Pillars.

13	10.0833	0.76	17506	....	14895	−2611	−15	13148	−4358	−25
10	7.5625	0.77	22948	....	21303	−1645	−7	18962	−3986	−18
15	7.5625	0.497	5262	....	5201	+29	+½	4681	−581	−11
8	3.7812	0.5	15107	....	11221	−3886	−25	9993	−5114	−34

FORMULAS FOR OBTAINING THE VOLUME AND WEIGHT OF A WROUGHT-IRON TIE UNDER A TENSILE STRAIN.

The strength of a wrought-iron tie under a strain of extension depends solely upon its minimum cross-section, its length being immaterial.

Assume a tie of a uniform cross-section of any shape whatever.

Let its length in feet be l.
Let its volume in cubic inches be V.
Let it be under the tensile force W.

Assuming that wrought-iron will just sustain a tensile force of 60,000 pounds per square inch, the number of square inches of section in the tie that sustains W is evidently $\frac{W}{60000}$. The length of the tie in inches is $12\,l$, and we, therefore, have for its volume,

$$V = \frac{W}{60000} \times 12\,l = \frac{Wl}{5000}$$

The weight of a cubic foot of wrought-iron is 480 pounds; hence $\frac{480}{1728} =$ the weight of a cubic inch of wrought-iron. Representing the weight of the tie by T, and multiplying the number of cubic inches in the tie by the weight of a cubic inch, we have

$$T = \frac{Wl \times 480}{5000 \times 1728} = \frac{40 \times Wl}{5000 \times 144}$$

$$T = \frac{Wl}{18000}$$

Which enables us to obtain directly the weight of a tie of known length that just breaks under a given strain.

We see from this that wrought-iron ties under the same tensile force vary in weight as their lengths, and that ties of the same length vary in weight as the tensile forces that act on them.

FORMULAS FOR OBTAINING THE STRENGTH, VOLUME, AND WEIGHT OF A CAST-IRON PILLAR UNDER A STRAIN OF COMPRESSION.

SOLID PILLARS WITH CIRCULAR SECTION AND ROUND ENDS.

Long Pillar.

Hodgkinson's formula for the strength of a solid cast-iron pillar with rounded ends and circular cross-section, whose length is not less than 15 times its diameter, and which is under a force of compression acting through its axis, is

$$(1) \quad W = 33379 \frac{d^{3.76}}{l^{1.7}}$$

in which

W is the weight in pounds which will just crush the pillar,
d the diameter of the pillar in *inches*,
l is the length of the pillar in *feet*.

The expression for the volume of the pillar in cubic inches is

$$(2) \quad V = \tfrac{1}{4} \pi d^2 \times 12\, l = 3 \pi d^2 l.$$

But from (1) we get

$$(3) \quad d^{3.76} = \frac{W l^{1.7}}{33379}$$

Whence $(4) \quad d = \left(\frac{W l^{1.7}}{33379}\right)^{\frac{1}{3.76}}$

And $(5) \quad d^2 = \left(\frac{W l^{1.7}}{33379}\right)^{\frac{2}{3.76}} = \left(\frac{W l^{1.7}}{33379}\right)^{\frac{1}{1.88}}$

Substituting this value of d^2 in (2) we get

$$(6) \quad V = 3 \pi l \left(\frac{W l^{1.7}}{33379}\right)^{\frac{1}{1.88}} = 3 \pi \left(\frac{1}{33379}\right)^{\frac{1}{1.88}} \times W^{\frac{1}{1.88}} \times l \times l^{\frac{1.7}{1.88}}.$$

$$(7) \quad V = 3 \pi \left(\frac{1}{33379}\right)^{\frac{1}{1.88}} \times W^{\frac{1}{1.88}} \times l^{\frac{1.79}{0.94}}$$

Reducing the constant terms, we have

Log. $33379 = 4.5234733$

Log. $\left(\frac{1}{33379}\right) = \bar{5}.4765267$

Log. $\left(\frac{1}{33379}\right)^{\frac{1}{1.88}} = \bar{5}.4765267 \div 1.88 = \bar{3}.5938972$

Log. $3 = 0.4771213$
Log. $\pi = 0.4971499$

Log. $0.036997151 = \bar{2}.5681684$

And (7) becomes

$$(8) \quad V = 0.036997151 \times W^{\frac{1}{1.9}} \times l^{\frac{1.79}{1.9}}$$

in which

V is the number of cubic inches in the pillar,
W is the weight that just crushes the pillar,
l is the length of the pillar in feet.

The weight of a cubic foot of cast-iron is 450 pounds. The weight of a cubic inch of cast-iron is therefore $\frac{450}{1728}$ pounds.

Multiplying the number of cubic inches in the pillar by $\frac{450}{1728}$, we get the weight of the pillar in pounds. Calling this P, we get

$$(9) \quad P = \frac{450}{1728} \times 0.036997151 \times W^{\frac{1}{1.9}} \times l^{\frac{1.79}{1.9}}.$$

Reducing the constant terms, we get

$$(10) \quad P = 0.009634674 \times W^{\frac{1}{1.9} = 0.5319148} \times l^{\frac{1.79}{0.9} = 1.9042553}$$

In which Log. $0.009634674 = \bar{3}.9838371$.

From (10) we see

1. When the weights to be sustained are equal, the weights of the pillars vary as the 1.9 powers of their lengths, or nearly as the squares of the lengths.

2. When the lengths of the pillars are equal, their weights vary as the 0.53 power of the weights to be sustained, or nearly as their square roots.

A rough formula which would very nearly give the number of pounds of iron in a long round-end pillar would be

$$(11) \quad P = 0.01 \sqrt{W} \times l^2 = \frac{l^2}{100} \sqrt{W}.$$

In bridges the quantities generally given are the breaking strain W, and the length l; and the quantity to be determined is the cross section or the diameter.

Equation (4) is

$$d = \left(\frac{W l^{1.7}}{33379}\right)^{\frac{1}{3.76}}$$

which expresses the relation sought.

It may be written

$$(4') \quad \log. d = \frac{\log. W + 1.7 \log. l - \log. 33379}{3.76}$$

—an equation which presents no difficulties.

Short Pillar.

The breaking weight of a short solid pillar with rounded ends is

$$W = 86237.5 \frac{d^2}{1 + 1.93769 \frac{l^{1.7}}{d^{1.76}}}$$

from which we get

$$\frac{W}{86237.5} = \frac{d^{2.76}}{d^{1.76} + 1.93769\, l^{1.7}}$$

$$\frac{W d^{1.76}}{86237.5} + \frac{1.93769}{86237.5} W l^{1.7} = d^{2.76}$$

$$(12)\quad d^{2.76} - \frac{1}{86237.5} W d^{1.76} = \frac{1.93769}{86237.5} W l^{1.7}$$

From which, by reducing the constants, we get

$$(13)\quad d^{2.76} - 0.0000115959\, W d^{1.76} = 0.0000224692\, W l^{1.7}$$

The value of d must be determined by trial. The first approximation will manifestly be obtained by neglecting the second term, and assuming for d a value greater than that thus determined.

To find the value of P,

P being the weight of the pillar, and V its volume in cubic inches, we have as before

$$P = \frac{450}{1728} V = \frac{450}{1728} \times 3\pi d^2 L$$

whence

$$d^2 = \frac{1728}{450 \times 3\pi l} P$$

$$d = \left(\frac{1728\, P}{450 \times 3\pi l}\right)^{\frac{1}{2}}$$

$$d^{1.76} = \left(\frac{1728\, P}{450 \times 3\pi l}\right)^{0.88} = \left(\frac{1728}{450 \times 3\pi}\right)^{0.88} \times \frac{P^{0.88}}{l^{0.88}}$$

$$d^{2.76} = \left(\frac{1728}{450 \times 3\pi}\right)^{1.38} \times \frac{P^{1.38}}{l^{1.38}}$$

Substituting in (12) we get

$$(14)\quad \left(\frac{1728}{450 \times 3\pi}\right)^{1.38} \times \frac{P^{1.38}}{l^{1.38}} - \frac{1}{86237.5} \left(\frac{1728}{450 \times 3\pi}\right)^{0.88} \times \frac{W P^{0.88}}{l^{0.88}} = \frac{1.93769}{86237.5} W l^{1.7}$$

whence, by a simple reduction,

$$(15)\quad P^{1.38} - 0.00002846057\, W l\, P^{0.88} = 0.00011909\, W l^{1.88}.$$

In this we get the approximate value of P by neglecting the second term, and afterwards increase the value of P so determined.

This equation may be written

$$(16) \quad P^{0.48}(P - 0.00002846057\, Wl) = 0.00011909\, Wl^{1.40},$$

which is probably a more convenient form for approximations.

SOLID PILLAR WITH CIRCULAR SECTION AND FLAT ENDS.

Long Pillar.

The formula for the breaking weight is

$$W = 98922 \frac{d^{2.55}}{l^{1.7}}$$

whence

$$d = \left(\frac{Wl^{1.7}}{98922}\right)^{\frac{1}{2.55}}$$

and

$$(17) \quad \log. d = \frac{\log. W + 1.7\log. l - \log. 98922}{3.55}$$

from which we can readily obtain d.

To obtain P,

We have, as before,

$$V = 3\pi d^2 l$$

whence

$$(18) \quad V = 3\pi l \left(\frac{Wl^{1.7}}{98922}\right)^{\frac{2}{2.55}} = \frac{1}{2.55}$$

$$(19) \quad V = 3\pi \left(\frac{1}{98922}\right)^{\frac{1}{1.775}} \times W^{\frac{1}{1.775}} \times l^{\frac{0.39}{1.775}}.$$

But

$$P = \frac{450}{1728} \times V.$$

hence

$$(20) \quad P = \frac{450}{1728} \times 3\pi \times \left(\frac{1}{98922}\right)^{\frac{1}{1.775}} \times W^{\frac{1}{1.775}} \times l^{\frac{1.39}{1.775}}$$

Reducing the constants, we get

$$(21) \quad P = 0.0037644 \times W^{\frac{1}{1.775}} \times l^{\frac{1.39}{1.775}}$$

which may be written logarithmically

$$(22) \quad \log. P = -2.424304 \times \frac{\log. W}{1.775} + \frac{1.39 \times \log. l}{0.71}$$

As 71 and 139 are prime numbers, it has been thought best not to reduce the exponent of l.

Short Pillar.

The formula is

$$W = 80237.5 \frac{d^2}{1 + 0.65383 \frac{l^{1.7}}{d^{2.55}}}$$

From which we get

$$(23) \quad \frac{W}{86237.5} = \frac{d^{2.55}}{d^{1.55} + 0.65383\, l^{1.7}}$$

$$(24) \quad d^{2.55} = \frac{W}{86237.5} d^{1.55} + \frac{0.65383}{86237.5} W l^{1.7}$$

$$(25) \quad d^{2.55} - 0.0000115950\, W d^{1.55} = 0.00000758173\, W l^{1.7}$$

which can only be solved by trial. As before, the first approximation to the value of d will be found by neglecting the second term.

To find P,

We have
$$P = \frac{450}{1728} V = \frac{450 \times 3\pi}{1728} d^2 l$$

whence
$$d = \left(\frac{1728\, P}{450 \times 3\pi\, l}\right)^{\frac{1}{2}}$$

$$d^{1.55} = \left(\frac{1728}{450 \times 3\pi}\right)^{0.775} \times \frac{P^{0.775}}{l^{0.775}}$$

$$d^{2.55} = \left(\frac{1728}{450 \times 3\pi}\right)^{1.275} \times \frac{P^{1.275}}{l^{1.275}}$$

Substituting these in (25) we get

$$(26) \quad \left(\frac{1728}{450 \times 3\pi}\right)^{1.275} \times \frac{P^{1.275}}{l^{1.275}} - \frac{W}{86237.5} \left(\frac{1728}{450 \times 3\pi}\right)^{0.775} \times \frac{P^{0.775}}{l^{0.775}} = \frac{0.65383}{86237.5} W l^{1.7}$$

whence

$$(27) \quad P^{1.275} - \frac{1}{86237.5}\left(\frac{450 \times 3\pi}{1728}\right) W l\, P^{0.775} = \left(\frac{450 \times 3\pi}{1728}\right)^{1.275} \times \frac{0.65383}{86237.5} W l^{2.416}$$

Reducing the constants, we get

$$(28) \quad P^{1.275} - 0.00002846057\, W l\, P^{0.775} = 0.00003731742\, W l^{2.416}.$$

Another equation that can only be solved by trial. This equation, like the similar one for short round-end pillars, may be written

$$(29) \quad P^{0.775}(P - 0.00002846057\, W l) = 0.00003731742\, W l^{2.416}$$

In either case of short pillars we may get P very readily, if d has been determined, from the equation

$$P = \frac{450}{1728} \times 3\pi d^2 l$$

$$(30) \quad P = 2.45437 \times d^2 \times l$$

HOLLOW PILLARS WITH CIRCULAR SECTION.

In examining Table III of Hodgkinson's experiments on hollow cast-iron pillars with circular section, we find that all of the pillars were of the same length, but of different weights and thicknesses. Assuming the formula which he deduced as sufficiently correct, let us examine the experiments themselves with a view to ascertain which of the pillars sustained the greatest weight in proportion to the quantity of iron it contained. We find as follows.

NO. OF PILLAR.	WEIGHT OF PILLAR.	BREAKING WEIGHT.	RATIO TO WEIGHT OF PILLAR.
14	59.5	40973	688
15	77.75	50477	649
13	50.25	26707	531

For all the other pillars the ratio of the breaking weight to the weight of the pillar was less than for the three given above. As the lengths of all the pillars were equal, we must conclude that in pillar No. 14 the material was used to the greatest advantage for obtaining the greatest sustaining power. It was apparently the best proportioned for sustaining the maximum weight with the minimum amount of material. In this pillar the exterior diameter was 3.36 inches, and the interior diameter 2.823. The difference between these, or the double thickness, is 0.537 inches, and the thickness is 0.2685 inches. Dividing the exterior diameter by the thickness we get 12.514 as their ratio. From this we conclude that as far as the experiments go they show that the most economical thickness for a hollow cast-iron pillar is a little less than one-twelfth of the exterior diameter. We will therefore assume that the best thickness for hollow cast-iron pillars is one-twelfth of the exterior diameter, and this thickness will be adopted in the calculations that follow.

MODIFIED FORMULAS FOR HOLLOW PILLARS (THICKNESS $\frac{1}{12}$).

Long, Round End.

The formula is
$$W = 29074 \frac{D^{3.76} - d^{3.76}}{l^{1.7}}$$

in which D is the exterior diameter, and d the interior diameter.

Assuming a thickness of one-twelfth, we have

(31) $\frac{1}{2}(D-d) = \frac{1}{12}D$
(32) $d = \frac{5}{6}D$
$d^{3.76} = (\frac{5}{6})^{3.76} \times D^{3.76} = 0.5038233 \, D^{3.76}$

Substituting this value of $d^{3.76}$ in the formula we get

(33) $W = 29074 \frac{D^{3.76} - 0.5038233 \, D^{3.76}}{l^{1.7}}$

whence

(34) $W = 14425.84 \times \frac{D^{3.76}}{l^{1.7}}$

Short, Round End.

Substituting $\frac{5}{6}D$ for d in the formula, we get

$$(35)\quad W = 86237.5 \left(\frac{D^2}{1 + 2.2246 \frac{l^{1.7}}{D^{1.76}}} - \frac{(\frac{5}{6})^2 D^2}{1 + 2.2246 \frac{l^{1.7}}{(\frac{5}{6})^{1.76} D^{1.76}}} \right)$$

Which reduces to

$$(36)\quad W = 86237.5 \left(\frac{D^2}{1 + 2.2246 \frac{l^{1.7}}{D^{1.76}}} - \frac{D^2}{1.44 + 4.41544 \frac{l^{1.7}}{D^{1.76}}} \right)$$

$$(37)\quad W = 86237.5\, D^2 \left(\frac{1}{1 + 2.2246 \frac{l^{1.7}}{D^{1.76}}} - \frac{1}{1.44 + 4.41544 \frac{l^{1.7}}{D^{1.76}}} \right)$$

Long, Flat End.

The formula is

$$W = 99318 \frac{D^{1.55} - d^{2.55}}{l^{1.7}}$$

Substituting $\frac{5}{6}D$ for d we get

$$(38)\quad W = 99318\, D^{2.55} \frac{[1 - (\frac{5}{6})^{2.55}]}{l^{1.7}}$$

$$99318\,[1 - (\frac{5}{6})^{2.55}] = 47326.25$$

Whence

$$(39)\quad W = 47326.25 \frac{D^{2.55}}{l^{1.7}}$$

Short, Flat End.

The formula is

$$W = 86237.5 \left(\frac{D^2}{1 + 0.651223 \frac{l^{1.7}}{D^{1.55}}} - \frac{d^2}{1 + 0.651223 \frac{l^{1.7}}{d^{1.55}}} \right)$$

Substituting $\frac{5}{6}D$ for d we get

$$(40)\quad W = 86237.5 \left(\frac{D^2}{1 + 0.651223 \frac{l^{1.7}}{D^{1.55}}} - \frac{(\frac{5}{6})^2 D^2}{1 + 0.651223 \frac{l^{1.7}}{(\frac{5}{6})^{1.55} D^{1.55}}} \right)$$

which becomes

$$(41)\quad W = 86237.5 \left(\frac{D^2}{1 + 0.651223 \frac{l^{1.7}}{D^{1.55}}} - \frac{D^2}{\frac{1}{(\frac{5}{6})^2} + \frac{0.651223}{(\frac{5}{6})^{1.55}} \times \frac{l^{1.7}}{D^{1.55}}} \right)$$

$$(42)\quad W = 86237.5\, D^2 \left(\frac{1}{1 + 0.651223 \frac{l^{1.7}}{D^{1.55}}} - \frac{1}{1.44 + 1.244007 \frac{l^{1.7}}{D^{1.55}}} \right)$$

HOLLOW PILLARS (*thickness* $\frac{1}{15}$ *exterior diameter.*)

To find Diameters and Weights.

Long, Round End.

From the formula just determined we get

$$D = \left(\frac{W l^{1.1}}{14425.84}\right)^{\frac{1}{5.16}}$$

Whence we get the logarithmic equation

(43) $\text{Log. } D = \dfrac{\log. W + 1.7 \log l - \log. 14425.84}{3.76}$

Which gives D when W and l are known.

To find P,

(44) $V = \tfrac{1}{4} \pi D^2 \times 12\, l - \tfrac{1}{4} \pi d^2 \times 12\, l = 3 \pi l\, (D^2 - d^2)$

Substituting $\tfrac{5}{6} D$ for d we have

(45) $V = 3 \pi l\, (D^2 - \dfrac{25}{36} D^2) = \dfrac{11}{12} \pi l\, D^2$

Substituting for D^2 its value as obtained from (34), and recollecting that

$$P = \frac{450}{1728} V$$

we have

(46) $P = \dfrac{450}{1728} \times \dfrac{11}{12} \pi l \left(\dfrac{W l^{1.1}}{14425.84}\right)^{\frac{1}{2.58}}$

(47) $P = \dfrac{450 \times 11 \pi}{1728 \times 12} W^{\frac{1}{2.58}} l^{\frac{3.58}{2.58}} \times \dfrac{1}{(14425.84)^{\frac{1}{2.58}}}$

(48) $P = \dfrac{450 \times 11 \pi}{1728 \times 12 \times (14425.84)^{\frac{1}{2.58}}} \times W^{\frac{1}{2.58}} l^{\frac{1.39}{2.58}}$

Reducing the constant,

(49) $P = 0.00459961 \times W^{\frac{1}{2.58}} \times l^{\frac{1.72}{2.58}}$

Short, Round End.

Transforming the formula we get

(50) $W = 86237.5\, D^2 \left(\dfrac{1.44 + 4.41544 \dfrac{l^{1.1}}{D^{1.76}} - 1 - 2.2246 \dfrac{l^{1.1}}{D^{1.76}}}{1.44 + 4.41544 \dfrac{l^{1.1}}{D^{1.76}} + 1.44 \times 2.2246 \dfrac{l^{1.1}}{D^{1.76}} + 4.41544 \times 2.2246 \dfrac{l^{2.2}}{D^{3.52}}}\right)$

(51) $W = 86237.5\, D^2 \left(\dfrac{0.44 + 2.19084 \dfrac{l^{1.1}}{D^{1.76}}}{1.44 + 7.61887 \dfrac{l^{1.1}}{D^{1.76}} + 9.8226 \dfrac{l^{2.2}}{D^{3.52}}}\right)$

IRON TRUSS BRIDGES FOR RAILROADS.

$$(52)\ W = 86237.5\, D^2 \left(\frac{0.44\, D^{3.62} + 2.19084\, l^{1.7}\, D^{1.76}}{1.44\, D^{5.62} + 7.61887\, l^{1.7}\, D^{1.76} + 9.8226\, l^{3.4}} \right)$$

$$(53)\ 1.44\, W D^{5.62} + 7.61887\, W\, l^{1.7}\, D^{1.76} + 9.8226\, W l^{3.4} = 37944.5 D^{5.62} + 188932.6\, l^{1.7} D^{3.76}$$

Transposing the terms containing D we have

$$(54)\ 37944.5 D^{5.62} + 188932.6\, l^{1.7} D^{3.76} - 1.44\, W D^{5.62} - 7.61887\, W\, l^{1.7} D^{1.76} = 9.8226\, W l^{3.4}$$

Dividing the coefficient of $D^{5.62}$ we get

$$(55)\ D^{5.62} + 4.979183\, l^{1.7} D^{2.76} - 0.00003795\, W D^{3.62} - 0.00020079\, W\, l^{1.7} D^{1.76} = 0.00025887\, W l^{3.4}$$

an equation which can only be solved by trial.

To find P,

P may be found after D has been determined, from the relation

$$(56)\ P = \frac{450}{1728} V = \frac{450\, \pi}{1728} \times \frac{11}{12} l D^2 = 0.7499464\, l D^2$$

Or the values of the different powers of D may be obtained from this equation and substituted in (55).

$$D^2 = \frac{P}{0.7499464\, l}$$

$$D^{1.76} = \frac{P^{0.88}}{(0.7499464)^{0.88}\, l^{0.88}}$$

$$D^{3.62} = \frac{P^{1.76}}{(0.7499464)^{1.76}\, l^{1.76}}$$

$$D^{2.76} = \frac{P^{1.88}}{(0.7499464)^{1.88}\, l^{1.88}}$$

$$D^{5.62} = \frac{P^{2.76}}{(0.7499464)^{2.76}\, l^{2.76}}$$

Substituting these values in (55) we get

$$(57)\ \frac{P^{2.76}}{(0.7499464)^{2.76}\, l^{2.76}} + \frac{4.979183\, l^{1.88}\, P^{1.88}}{(0.7499464)^{1.88}\, l^{1.88}} - \frac{0.00003795\, W P^{1.76}}{(0.7499464)^{1.76}\, l^{1.76}} - \frac{0.00020079\, W l^{0.82} P^{0.88}}{(0.7499464)^{0.88}} = 0.00025887\, W l^{3.4}$$

Multiplying by $(0.7499464)^{2.76}\, l^{2.76}$

we get

$$(58)\ P^{2.76} + 4.979183\, (0.7499464)^{0.88}\, l^{0.88} P^{1.88} - 0.00003795 \times 0.7499464\, W l\, P^{1.76} - 0.00020079\, (0.7499464)^{1.88}\, W l^{3.64} P^{0.88} = 0.00025887\, (0.7499464)^{2.76}\, W l^{6.16}$$

Which by reduction of the constants becomes

$$(59)\ P^{2.76} + 3.865312\, l^{0.88} P^{1.88} - 0.00002846057\, W l\, P^{1.76} - 0.0001168956\, W l^{3.64} P^{0.88} = 0.0001169933\, W l^{6.16}.$$

Which is another equation which can only be solved by trial.

Long, Flat End.

The formula is
$$W = 47326.25 \frac{D^{2.45}}{l^{1.7}}$$

Whence

$$(60) \quad D = \left(\frac{W l^{1.7}}{47326.25}\right)^{\frac{1}{2.45}}$$

From which we deduce the logarithmic equation

$$(61) \quad \text{Log. } D \frac{\log. W + 1.7 \log. l - \log. 47326.25}{3.55}$$

Whence we determine D when W and l are known.

To find P:

By the same process as in long round-end hollow pillars we get the following equation.

$$(62) \quad P = \frac{450}{1728} \times \frac{11}{12} \pi l \left(\frac{W l^{1.7}}{47326.25}\right)^{\frac{2}{2.45}}$$

$$(63) \quad P = \frac{450 \times 11 \pi}{1728 \times 12 \times (47326.25)^{\frac{2}{2.45}}} \times W^{\frac{2}{2.45}} \times l^{\frac{5.45}{2.45}}$$

Reducing the constants we get

$$(64) \quad P = 0.00174240 \times W^{\frac{2}{2.55}} \times l^{\frac{1.11}{2.11}}$$

An equation easily solved by logarithms. The fractional exponents can be more readily used as they are than if they were reduced.

Short, Flat End.

The formula for the breaking weight is:

$$(65) \quad W = 86237.5 \, D^2 \left(\frac{1}{1 + 0.651223 \frac{l^{1.7}}{D^{1.75}}} - \frac{1}{1.44 + 1.244007 \frac{l^{1.7}}{D^{1.85}}}\right)$$

$$(66) \quad W = 86237.5 \, D^2 \left(\frac{D^{1.55}}{D^{1.55} + 0.651223 \, l^{1.7}} - \frac{D^{1.85}}{1.44 \, D^{1.85} + 1.244007 \, l^{1.7}}\right)$$

$$(67) \quad W = 86237.5 \, D^2 \left(\frac{1.44 \, D^{3.1} + 1.244007 \, l^{1.7} \, D^{1.55} - D^{3.1} - 0.651223 \, l^{1.7} \, D^{1.86}}{1.44 \, D^{3.1} + 1.244007 \, l^{1.7} \, D^{1.55} + 1.44 \times 0.651223 \, l^{1.7} \, D^{1.85} + 0.651223 \times 1.244007 \, l^{3.4}}\right)$$

$$(68) \quad W = 86237.5 \, D^2 \left(\frac{0.44 \, D^{3.1} + 0.592784 \, l^{1.7} \, D^{1.55}}{1.44 \, D^{3.1} + (1.244007 + 1.44 \times 0.651223) \, l^{1.7} \, D^{1.85} + 0.651223 \times 1.244007 \, l^{3.4}}\right)$$

$$(69) \quad W = 86237.5 \, D^2 \left(\frac{0.44 \, D^{3.1} + 0.592784 \, l^{1.7} \, D^{1.55}}{1.44 \, D^{3.1} + 2.181767 \, l^{1.7} \, D^{1.85} + 0.810125 \, l^{3.4}}\right)$$

$$(70) \quad 1.44 \, W D^{3.1} + 2.181767 \, W l^{1.7} \, D^{1.55} + 0.810125 \, W l^{3.4} = 37944.5 \, D^{5.1} + 51120.2 \, l^{1.7} \, D^{3.55}$$

Transposing the terms containing D, we have

$$(71) \quad 37944.5 \, D^{5.1} + 51120.2 \, l^{1.7} \, D^{3.55} - 1.44 \, W D^{3.1} - 2.181767 \, W l^{1.7} \, D^{1.55} = 0.810125 \, W l^{3.4}.$$

IRON TRUSS BRIDGES FOR RAILROADS.   47

Dividing by the coefficient of $D^{5.1}$, we get

(72) $D^{5.1} + 1.347236\, l^{1.1} D^{3.45} - 0.00003795\, W D^{3.1} - 0.0000575\, W l^{1.1} D^{1.55} = 0.00002135\, W l^{2.4}$.

An equation similar to the one deduced for short, round-end, hollow pillars.

To find P:
We have, as before, (56),
$$P = 0.7499464\, l\, D^2$$
Whence
$$D = \left(\frac{P}{0.7499464\, l}\right)^{\frac{1}{2}}$$
$$D^{1.55} = \left(\frac{P}{0.7499464\, l}\right)^{0.775}$$
$$D^{3.1} = \left(\frac{P}{0.7499464\, l}\right)^{1.55}$$
$$D^{3.45} = \left(\frac{P}{0.749964\, l}\right)^{1.775}$$
$$D^{5.1} = \left(\frac{P}{0.7499464\, l}\right)^{2.55}$$

Substituting these values in (72) we get

(73) $\dfrac{P^{2.55}}{(0.7499464)^{2.55} l^{2.55}} + \dfrac{1.347236\, P^{1.775}}{(0.7499464)^{1.775} l^{0.675}} - \dfrac{0.00003795\, W P^{1.55}}{(0.7499464)^{1.55} l^{1.55}} - \dfrac{0.0000575\, W l^{0.075} P^{0.775}}{(0.7499464)^{0.775}}$
$= 0.00002135\, W l^{2.4}$.

Clearing fractions, we get

(74) $P^{2.55} + 1.347236\, (0.7499464)^{0.775} l^{2.475} P^{1.775} - 0.00003795 \times 0.7499464\, W l\, P^{1.55}$
$- 0.0000575\, (0.7499464)^{1.775} W l^{2.475} P^{0.775} = 0.00002135\, (0.7499464)^{2.55} W l^{4.95}$.

Which reduces to

(75) $P^{2.55} + 1.0779333\, l^{2.475} P^{1.775} - 0.00002846057\, W l\, P^{1.55} - 0.0000345015\, W l^{2.475} P^{0.775} = 0.00001025015\, W l^{4.95}$.

Another equation to be solved by approximations.

Comparing together the formulas for the weights of long solid pillars and long hollow ones, we see that they differ only in the coefficients, which are to each other as 2.1 is to 1 —in other words, that for the same length and the same strain a solid pillar needs more than twice as much metal as a hollow one.

We will now gather together all the formulas which have been deduced for the strength of a pillar, and those that enable us to find its diameter or its weight.

FORMULAS.

$W =$ breaking weight of the pillar in pounds.
$P \;=$ weight of the pillar in pounds.
$l \;\;=$ length of the pillar in feet.
$d \;=$ diameters of the pillar in inches.
$D \;=$ exterior diameter of a hollow pillar in inches.
LONG *round-end* pillars are over 15 diameters in length.
LONG *flat-end* pillars are over 30 diameters in length.

Ends.	Length.	Formula.	Logarithms.	
		Solid.		
Round,	Long,	$W = 33379 \dfrac{d^{2.76}}{l^{1.7}}$	log. 33379	$= 4.5234733$
Round,	Short,	$W = 86237.5 \dfrac{d^2}{1 + 1.93769 \dfrac{l^{1.7}}{d^{1.76}}}$	log. 86237.5 log. 1.93769	$= 4.9356962$ $= 0.2872641$
Flat,	Long,	$W = 98922 \dfrac{d^{2.88}}{l^{1.7}}$	log. 98922	$= 4.9952929$
Flat,	Short,	$W = 86237.5 \dfrac{d^2}{1 + 0.65383 \dfrac{l^{1.7}}{d^{1.88}}}$	log. 86237.5 log. 0.65383	$= 4.9356962$ $= \bar{1}.8154645$
		Hollow (any thickness).		
Round,	Long,	$W = 29074 \dfrac{D^{2.75} - d^{2.76}}{l^{1.7}}$	log. 29074	$= 4.4635048$
Round,	Short,	$W = 86237.5 \left(\dfrac{D^2}{1 + 2.2246 \dfrac{l^{1.7}}{D^{1.76}}} - \dfrac{d^2}{1 + 2.2246 \dfrac{l^{1.7}}{d^{1.76}}} \right)$	log. 86237.5 log. 2.2246	$= 4.9356962$ $= 0.3472526$
Flat,	Long,	$W = 99318 \dfrac{D^{2.88} - d^{2.88}}{l^{1.7}}$	log. 99318	$= 4.9970280$
Flat,	Short,	$W = 86237.5 \left(\dfrac{D^2}{1 + 0.651223 \dfrac{l^{1.7}}{D^{1.88}}} - \dfrac{d^2}{1 + 0.651223 \dfrac{l^{1.7}}{d^{1.88}}} \right)$	log. 86237.5 log. 0.651223	$= 4.9356962$ $= \bar{1}.8137294$
		Hollow (thickness $\tfrac{1}{11}$).		
Round,	Long,	$W = 14425.84 \dfrac{D^{2.76}}{l^{1.7}}$	log. 14425.84	$= 4.1591411$
Round,	Short,	$W = 86237.5 \, D^2 \left(\dfrac{1}{1 + 2.2246 \dfrac{l^{1.7}}{D^{1.76}}} - \dfrac{1}{1.44 + 4.41544 \dfrac{l^{1.7}}{D^{1.76}}} \right)$	log. 86237.5 log. 2.2246 log. 1.44 log. 4.41544	$= 4.9356962$ $= 0.3472526$ $= 0.1583625$ $= 0.6449741$
Flat,	Long,	$W = 47326.25 \dfrac{D^{2.88}}{l^{1.7}}$	log. 47326.25	$= 4.6751021$
Flat,	Short,	$W = 86237.5 \, D^2 \left(\dfrac{1}{1 + 0.651223 \dfrac{l^{1.7}}{D^{1.88}}} - \dfrac{1}{1.44 + 1.244007 \dfrac{l^{1.7}}{D^{1.88}}} \right)$	log. 86237.5 log. 0.651223 log. 1.44 log. 1.244007	$= 4.9356962$ $= \bar{1}.8137294$ $= 0.1583625$ $= 0.0948228$

IRON TRUSS BRIDGES FOR RAILROADS.

ENDS.	LENGTH.	FORMULA.	LOGARITHMS.
Round,	Long,	**Solid.** $$\text{Log. } d = \frac{\log. W + 1.7 \log. l - 33379}{3.76}$$ $P = 2.45437\, l\, d^3$ $P = 0.009634674\, W^{\frac{1}{1.85}} l^{\frac{1.79}{5.94}}$	log. 33379 = 4.5234733 log. 2.45437 = 0.3899399 log. 0.009634674 = $\bar{3}$.9838371
Round,	Short,	$d^{2.76} - 0.0000115959\, W\, d^{1.16} = 0.0000224692\, W\, l^{1.7}$ $P = 2.45437\, l\, d^2$ $P^{1.68} - 0.00002846057\, W\, l\, P^{1.10} = 0.00011909\, W\, l^{2.68}$	log. 0.0000115959 = $\bar{5}$.0643038 log. 0.0000224692 = $\bar{5}$.3515879 log. 2.45437 = 0.3899399 log. 0.00002846057 = $\bar{5}$.4542437 log. 0.00011909 = $\bar{4}$.0758749
Flat,	Long,	$$\text{Log. } d = \frac{\log. W + 1.7 \log. l - \log. 98922}{3.55}$$ $P = 2.45437\, l\, d^2$ $P = 0.0037644\, W^{\frac{1}{1.775}} l^{\frac{1.79}{5.71}}$	log. 98022 = $\bar{4}$.9952929 log. 2.45437 = 0.3899399 log. 0.0037644 = $\bar{3}$.5756960
Flat,	Short,	$d^{2.55} - 0.0000115959\, W\, d^{1.16} = 0.00000758173\, W\, l^{1.7}$ $P = 2.45437\, l\, d^2$ $P^{1.176} - 0.00002846057\, W\, l\, P^{0.776} = 0.00003731742\, W\, l^{2.475}$	log. 0.0000115959 = $\bar{5}$.0643038 log. 0.00000758173 = $\bar{6}$.8797683 log. 2.45437 = 0.3899399 log. 0.00002846057 = $\bar{5}$.4542437 log. 0.00003731742 = $\bar{5}$.5719116
Round,	Long,	**Hollow** (*thickness* $\tfrac{1}{12}$). $$\text{Log. } D = \frac{\log. W + 1.7 \log. l - \log. 14425.84}{3.76}$$ $P = 0.7499464\, l\, D^2$ $P = 0.00459961 \times W^{\frac{1}{1.88}} l^{\frac{1.79}{5.94}}$	log. 14425.84 = 4.1591411 log. 0.7499464 = $\bar{1}$.8750301 log. 0.00459961 = $\bar{3}$.6627210
Round,	Short,	$D^{3.12} + 4.979183\, l^{1.7} D^{3.16} - 0.00003795\, W D^{3.52} - 0.00020079\, W\, l^{1.7} D^{1.76} = 0.00025887\, W\, l^{3.4}$ $P = 0.7499464\, l\, D^2$ $P^{2.76} + 3.865312\, l^{2.58} P^{1.89} - 0.00002846057\, W\, l\, P^{1.10}\, 0.0001168956\, W\, l^{3.49} P^{0.68} = 0.0001169033\, W\, l^{5.16}$	log. 4.979183 = 0.6971580 log. 0.00003795 = $\bar{5}$.5792136 log. 0.00020079 = $\bar{4}$.3027417 log. 0.00025887 = $\bar{4}$.4130778 log. 0.7499464 = $\bar{1}$.8750301 log. 3.865312 = 0.5871845 log. 0.00002846057 = $\bar{5}$.4542437 log. 0.0001168956 = $\bar{4}$.0677983 log. 0.0001169033 = $\bar{4}$.0681609
Flat,	Long,	$$\text{Log. } D = \frac{\log. W + 1.7 \log. l - \log. 47326.25}{3.55}$$ $P = 0.7499464\, l\, D^2$ $P = 0.00174249 \times W^{\frac{2}{1.68}} l^{\frac{1.41}{5.71}}$	log. 47326.25 = 4.6751021 log. 0.7499464 = $\bar{1}$.8750301 log. 0.00174249 = $\bar{3}$.2411698

Ends.	Length.	Formula.	Logarithms.
Flat,	Short,	*Hollow* (*thickness*, $\frac{1}{12}$)—continued. $D^{1.8} + 1.347236\, l^{1.7} D^{2.15} - 0.00003795\, WD^{2.1} - 0.0000575$ $Wl^{1.7} D^{1.55} = 0.00002135\, Wl^{2.4}$ $P = 0.7499464\, l\, D^2$ $P^{2.55} + 1.0779333\, l^{2.475} P^{1.775} - 0.00002846057\, Wl\, P^{1.66}$ $- 0.0000345015\, Wl^{2.475} P^{0.775} = 0.00001025015\, Wl^{5.25}$	log. 1.347236 = 0.1294437 log. 0.00003795 = $\overline{5}$.5792136 log. 0.0000575 = $\overline{5}$.7596594 log. 0.00002135 = $\overline{5}$.3294033 log. 0.7499464 = $\overline{1}$.8750301 log. 1.0779333 = 0.0325920 log. 0.00002846057 = $\overline{5}$.4542439 log. 0.0000345015 = $\overline{5}$.5378378 log. 0.00001025015 = $\overline{5}$.0107301

In using these formulas in the comparison that follows we will adopt equation (49), which is for *long round-end hollow pillars, with a thickness one-twelfth of the exterior diameter*. For perfect accuracy we should use the short-pillar formula for the segments of the Top Chord, but the method adopted is believed to be sufficiently accurate in a comparison, and it saves a great deal of tedious computation.

For flat-end pillars, not immovably fastened at their extremities, we might use the flat-end formulas by suitably modifying W. It would probably be accurate for segments of the Top Chord to deduce their breaking weights from the flat-end formulas and divide by 2 instead of 3. There must be some intermediate state between flat-end pillars perfectly fixed, and those that are perfectly free. The latter have one-third the strength of the former, while a pillar that can only be subjected to very small motions, like a Top Segment, would probably have an intermediate strength, requiring a divisor of 2 or 1½.

CONSTANTS USED IN CALCULATING WEIGHTS.

The formula for calculating the weight of a long hollow cast-iron pillar, with circular section and thickness $\frac{1}{12}$, that just breaks under the strain W, is

$$(1) \quad P = 0.00459961 \times W^{\frac{1}{1.58}} \times l^{\frac{1.79}{6.91}}.$$

In applying this we must take for W five times the actual maximum strain on the pillar, as that is the least strain under which the pillar is expected to break. Therefore, if we take for W the strain determined by calculation, we must write $5W$ for W in the formula. Hence,

$$(2) \quad P = 0.00459961 \times (5W)^{\frac{1}{1.58}} \times l^{\frac{1.79}{6.91}}.$$

$$(3) \quad P = 0.00459961 \times (5)^{\frac{1}{1.58}} \times l^{\frac{1.79}{6.91}} W^{\frac{1}{1.58}}.$$

The first two factors of the second number of this formula are constant, and the third term has but five different values in the trusses which we are to examine. It will therefore much simplify our calculations if we first find the logarithm of the product of the first three terms for these different values of l. It will then, for each particular case, only be necessary to add to the proper one of these logarithms the logarithm of $W^{\frac{1}{1.58}}$, and their sum will be the logarithm of P. These logarithms are

$l = \quad\quad = 12.5 \ldots$ log. $(0.00459961 \times 5^{\frac{1}{1.58}} \times l^{\frac{1.79}{6.91}}) = 0.1233103$

$l = \quad\quad = 18.75 \ldots$ log. $(0.00459961 \times 5^{\frac{1}{1.58}} \times l^{\frac{1.79}{6.91}}) = 0.4586331$

$l = 18.75 \times \tfrac{1}{3}\sqrt{10} = 19.76 \ldots$ log. $(0.00459961 \times 5^{\frac{1}{1.58}} \times l^{\frac{1.79}{6.91}}) = 0.5022000$

$l = 18.75 \times \tfrac{1}{3}\sqrt{13} = 22.53 \ldots$ log. $(0.00459961 \times 5^{\frac{1}{1.58}} \times l^{\frac{1.79}{6.91}}) = 0.6106886$

$l = \quad\quad = 25. \ldots$ log. $(0.00459961 \times 5^{\frac{1}{1.58}} \times l^{\frac{1.79}{6.91}}) = 0.6965483$

In using which we take for W the maximum calculated compression on the member in question.

The formula for calculating the weight of a wrought iron member of uniform section, that is under a tensile strain, is

$$(1) \quad T = \frac{Wl}{18000}$$

In which W, as before, is the breaking weight. As we estimate for a breaking weight 5 times as great as the greatest strain that can come on the part, if we take W as the strain found by calculation, we must write $5W$ for W in the formula. Hence,

$$(2) \quad T = \frac{5Wl}{18000}$$

or

$$(3) \quad T = \frac{l}{3600} \times W$$

But the first factor of the second member has but few values. Determining the logarithms of these values beforehand, we have but to add to the proper logarithm the logarithm of W and we at once get the logarithm of T. These constants are,

$$l = \phantom{18.75 \times \tfrac{1}{3}\sqrt{13}} = 12.5 \ldots \log.\left(\frac{l}{3600}\right) = \overline{3}.5406075$$

$$l = \phantom{18.75 \times \tfrac{1}{3}\sqrt{13}} = 18.75 \ldots \log.\left(\frac{l}{3600}\right) = \overline{3}.7166968$$

$$l = 18.75 \times \tfrac{1}{3}\sqrt{10} = 19.76 \ldots \log.\left(\frac{l}{3600}\right) = \overline{3}.7395775$$

$$l = 18.75 \times \tfrac{1}{3}\sqrt{13} = 22.53 \ldots \log.\left(\frac{l}{3600}\right) = \overline{3}.7965492$$

$$l = \phantom{18.75 \times \tfrac{1}{3}\sqrt{13}} = 25 \ldots \log.\left(\frac{l}{3600}\right) = \overline{3}.8416375$$

$$l = 18.75 \times \phantom{\tfrac{1}{3}} \sqrt{2} = 26.5 \ldots \log.\left(\frac{l}{3600}\right) = \overline{3}.8092138$$

$$l = 25 \times \tfrac{1}{2}\sqrt{5} = 27.95 \ldots \log.\left(\frac{l}{3600}\right) = \overline{3}.8900925$$

$$l = 25 \times \phantom{\tfrac{1}{2}}\sqrt{2} = 35.36 \ldots \log.\left(\frac{l}{3600}\right) = \overline{3}.9921525$$

W is the calculated maximum strain.

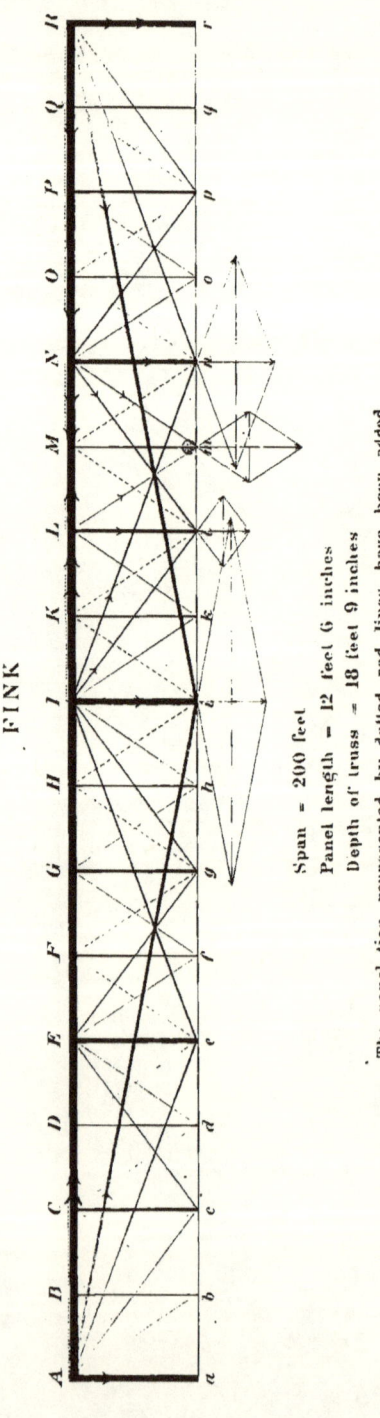

FINK

Span = 200 feet
Panel length = 12 feet 6 inches
Depth of truss = 18 feet 9 inches

The panel ties represented by dotted red lines have been added

THE FINK TRUSS.

This is hardly a truss, being more properly a trussed girder. It has no Bottom Chord; the line $a, b, c, \ldots\ldots p, q, r$, being merely a line of horizontal posts to keep the points of support at their proper relative distances apart. The theory of construction is, first to support the middle point by a post which itself is held up by ties extending from its lower extremity to the ends of the Top Chord. The middle point is then considered as fixed, and the halves of the Top Chord are bisected, the points of section being upheld by posts that are supported by ties that extend to the middle point, and to one of the ends of the Top Chord. These quarter points are then considered as fixed, and the Top Chord is divided into eighths, that are similarly upheld by posts and ties to the nearest fixed points. The Top Chord is again subdivided into sixteenths that are similarly supported. The number of subdivisions depends upon the span of the truss, and the length of panel that may be desired. If, in the case which we have chosen, it had been thought best to have but twelve panels, the distance IN would have been trisected, and the lower extremities of the posts would have been connected by ties with I and N.

Assumed Weights and Dimensions.

$$ar \ldots\ldots\ldots\ldots\ldots\ldots\ldots\ldots\ldots\ldots\ldots\ldots\ldots\ldots\ldots = 200'$$
$$\text{No. of panels} \ldots\ldots\ldots\ldots\ldots\ldots\ldots\ldots\ldots\ldots\ldots = 16$$
$$\text{Panel length} \ldots\ldots\ldots\ldots\ldots\ldots\ldots\ldots\ldots\ldots\ldots = 12' \, 6''$$
$$\text{Depth of truss} \ldots\ldots\ldots\ldots\ldots\ldots\ldots\ldots\ldots\ldots\ldots = 18' \, 9''$$
$$\text{Panel weight of engine} \ldots\ldots\ldots\ldots\ldots\ldots\ldots\ldots = 17600 = w' + e$$
$$\text{``\qquad ``\qquad tender} \ldots\ldots\ldots\ldots\ldots\ldots\ldots\ldots = 16160 = w' + t$$
$$\text{``\qquad ``\qquad cars} \ldots\ldots\ldots\ldots\ldots\ldots\ldots\ldots\ldots = 13152 = w'$$
$$\text{``\qquad ``\qquad bridge} \ldots\ldots\ldots\ldots\ldots\ldots\ldots\ldots = 9375 = w''$$
$$\text{Excess of panel weight of engine over panel weight of cars} .. = 4448 = e$$
$$\text{``\qquad ``\qquad ``\qquad tender \quad ``\qquad ``\qquad ``\qquad ``} \ldots = 3008 = t$$

Transmission of Strains.

To show how the weights attached to this truss are transmitted through the different parts to the abutments, we will assume a weight at m, and trace its effect on the truss. It is first halved, and the halves are carried by mL and mN to L and N. The half at N is carried by the post Nn to n, and thence a quarter of the original weight goes to R, and a quarter to I. The quarter at I goes through the part Ii to i, whence one-eighth goes to R and one-eighth to A. Returning to the half of the weight that went to L, we find that from l one-quarter goes to I, and one-quarter to N. The quarter at

I is halved at i, and one-eighth goes to R and one-eighth to A. The quarter at N becomes one-eighth at R and one-eighth at I. The one-eighth at I becomes one-sixteenth at A and one-sixteenth at R. All the fractions of the original weight at m have now reached the abutments. Gathering together the parts that have reached A and R we find at A $\frac{1}{2}+\frac{1}{8}+\frac{1}{16}=\frac{6}{16}$, and at R $\frac{1}{4}+\frac{1}{8}+\frac{1}{8}+\frac{1}{8}+\frac{1}{16}=\frac{11}{16}$. As the point m is five panel lengths from r, and eleven panel lengths from a, the proportions of the original weight which we have carried to the two abutments are evidently those that theory would have told us in advance should have gone there. We may therefore conclude that the resolution was correctly made. The strain on each tie will be equal to the *vertical* weight which it carries, multiplied by the secant of its inclination with the vertical.

Were there a Bottom Chord our resolution would have been made along the theoretical lines mR and mA. Were the combination to consist only of the Top Chord, and the ties mL and mN, still the same proportions of the original weight would ultimately have reached A and R, though, after getting to L and N by the ties, they would have been carried to the abutments by the Top acting as a solid beam.

From Bridge and Moving Load.

The principle of the counterbalancing of equal weights similarly situated, which is so important in trusses with Bottom Chords, does not come into use in this bridge, and the strains from the Moving Load, and from the Weight of the Bridge may be considered together in calculation.

Maxima Direct Compressions on Posts.

There are no transmitted strains on the posts Kk, Mm, Oo, and Qq, and the only direct strains which they have to sustain are those due to the proper proportions of the weight of the segments of the Top Chord, of the ties, and of the top lateral bracing. These are parts of the permanent bridge weight, no direct strain from the moving load coming on the posts in question. But as we have agreed for all trusses to consider in this preliminary calculation the parts as without weight, and the weight of the truss as represented by weights suspended along the Bottom Chord at each roadway bearer, we will find no direct compressions on the posts mentioned above. Examining the effect of the train, which comes on from a, with the head of the train at different points, we find the following positions give the maxima strains on the different posts:

on Ii ...head of train at q
on Ll.... " " m
on Nn.... " " q
on Pp.... " " q
on Rr.... " " q

We, therefore, have on the posts:

on Ii .... $7(w'+w'') + \frac{3}{8}c + \frac{7}{8}t = 161989$
on Kk 0
on Ll .... $(w'+w'') + \frac{1}{2}c + \frac{1}{2}t = 26255$
on Mm 0
on Nn .... $3(w'+w'') + \frac{3}{4}e + \frac{3}{4}t = 73173$
on Oo 0
on Pp .... $(w'+w'') + \frac{1}{2}c + \frac{1}{2}t = 26255$
on Qq 0
on Rr .... $\frac{15}{2}(w'+w'') + \frac{23}{16}e + \frac{7}{16}t = 181715$

Maxima Direct Tensions on Ties.

In examining the strains on the ties we can readily see that each pair of ties receives its maximum strain when the post that rests on them is under its maximum compression. But the compression on the post, being a vertical force, must be multiplied by the secant of the inclination of the tie to the vertical, to get the strain on the tie. Moreover, this vertical force is increased by the weight on the roadway bearer at the intersection of the ties. We, therefore, have on the ties:

on kI .... $\frac{1}{2}(w'+w''+e) \times \frac{1}{3}\sqrt{13} = 16210$
on kL .... $\frac{1}{2}(w'+w''+e) \times \frac{1}{3}\sqrt{13} = 16210$
on mL .... $\frac{1}{2}(w'+w''+e) \times \frac{1}{3}\sqrt{13} = 16210$
on mN .... $\frac{1}{2}(w'+w''+e) \times \frac{1}{3}\sqrt{13} = 16210$
on oN .... $\frac{1}{2}(w'+w''+e) \times \frac{1}{3}\sqrt{13} = 16210$
on oP .... $\frac{1}{2}(w'+w''+e) \times \frac{1}{3}\sqrt{13} = 16210$
on qP .... $\frac{1}{2}(w'+w''+e) \times \frac{1}{3}\sqrt{13} = 16210$
on qR .... $\frac{1}{2}(w'+w''+e) \times \frac{1}{3}\sqrt{13} = 16210$
on lI .... $\frac{1}{2}[26255 + (w'+w''+e)] \times \frac{1}{3}\sqrt{25} = 44358$
on lN .... $\frac{1}{2}[26255 + (w'+w''+e)] \times \frac{1}{3}\sqrt{25} = 44358$
on pN .... $\frac{1}{2}[26255 + (w'+w''+e)] \times \frac{1}{3}\sqrt{25} = 44358$
on pR .... $\frac{1}{2}[26255 + (w'+w''+e)] \times \frac{1}{3}\sqrt{25} = 44358$
on nI .... $\frac{1}{2}[73173 + (w'+w''+t)] \times \frac{1}{3}\sqrt{73} = 140560$
on nR .... $\frac{1}{2}[73173 + (w'+w''+t)] \times \frac{1}{3}\sqrt{73} = 140560$
on iR .... $\frac{1}{2}[161989 + (w'+w'')] \times \frac{1}{3}\sqrt{265} = 500617$

Maxima Compressions on Top Chord.

The greatest compression on the Top Chord will evidently be developed when there is the greatest load upon the bridge. This, of course, will be when the head of the train is at q. We will then have $w'+e$ at p and q, $w'+t$ at n and o, and w' at all the other points of support; and w'', of course, at all the points. Each weight affecting the truss independently of the others, and there being no counter-balancing of equal weights, we have evidently merely to determine the compressions due to each weight, and add them together. These compressions are as follows:

From weight at b — on RI $\frac{1}{8}$ $(w' + w'')$
on IE $\frac{3}{8}$ $(w' + w'')$
on EC $\frac{3}{8}$ $(w' + w'')$
on CA $\frac{3}{8}$ $(w' + w'')$
From weight at c — on RI $\frac{3}{8}$ $(w' + w'')$
on IE $\frac{3}{8}$ $(w' + w'')$
on EA $\frac{5}{8}$ $(w' + w'')$
From weight at d — on RI $\frac{3}{8}$ $(w' + w'')$
on IE $\frac{5}{8}$ $(w' + w'')$
on EC $\frac{5}{8}$ $(w' + w'')$
on CA $\frac{3}{8}$ $(w' + w'')$
From weight at e — on RI $\frac{3}{8}$ $(w' + w'')$
on IA $\frac{5}{8}$ $(w' + w'')$
From weight at f — on RI $\frac{5}{8}$ $(w' + w'')$
on IG $\frac{5}{8}$ $(w' + w'')$
on GE $1\frac{3}{8}$ $(w' + w'')$
on EA $\frac{5}{8}$ $(w' + w'')$
From weight at g — on RI $\frac{5}{8}$ $(w' + w'')$
on IE $1\frac{3}{8}$ $(w' + w'')$
on EA $\frac{5}{8}$ $(w' + w'')$
From weight at h — on RI $\frac{3}{8}$ $(w' + w'')$
on IG $1\frac{3}{8}$ $(w' + w'')$
on GE $\frac{5}{8}$ $(w' + w'')$
on EA $\frac{3}{8}$ $(w' + w'')$
From weight at i — on RA $\frac{5}{8}$ $(w' + w'')$
From weight at k — on RN $\frac{3}{8}$ $(w' + w'')$
on NL $\frac{3}{8}$ $(w' + w'')$
on LI $1\frac{3}{8}$ $(w' + w'')$
on IA $\frac{3}{8}$ $(w' + w'')$
From weight at l — on RN $\frac{3}{8}$ $(w' + w'')$
on NI $1\frac{3}{8}$ $(w' + w'')$
on IA $\frac{3}{8}$ $(w' + w'')$
From weight at m — on RN $\frac{3}{8}$ $(w' + w'')$
on NL $1\frac{3}{8}$ $(w' + w'')$
on LI $\frac{3}{8}$ $(w' + w'')$
on IA $\frac{3}{8}$ $(w' + w'')$
From weight at n — on RI $\frac{1}{8}$ $(w' + w'')$
on IA $\frac{3}{8}$ $(w' + w'')$
From weight at o — on RP $\frac{1}{2}$ $(w' + t + w'')$
on PN $\frac{1}{2}$ $(w' + t + w'')$
on NI $\frac{3}{2}$ $(w' + t + w'')$
on IA $\frac{3}{2}$ $(w' + t + w'')$
From weight at p — on RN $\frac{3}{8}$ $(w' + e + w'')$
on NI $\frac{3}{8}$ $(w' + e + w'')$
on IA $\frac{3}{8}$ $(w' + e + w'')$
From weight at q — on RP $\frac{1}{2}$ $(w' + e + w'')$
on PN $\frac{3}{2}$ $(w' + e + w'')$
on NI $\frac{3}{2}$ $(w' + e + w'')$
on IA $\frac{1}{2}$ $(w' + e + w'')$

Adding together these different independent strains, we obtain the following maxima strains on the Top Chord :

on $A\ I\ \frac{2A}{7}\ (w' + w'') + \frac{3}{7}c + \frac{3}{7}t = 649731.33$
on $I\ L\ \frac{2A}{7}\ (w' + w'')\frac{6}{7} + \frac{6}{7}c + \frac{1}{7}M = 661198.$
on $L\ N\ \frac{2A}{7}\ (w' + w'') \cdot \frac{6}{7}c + \frac{1}{7}M = 661198.$
on $N\ P\ \frac{2A}{7}\ (w' + w'') + \frac{5}{7}c + \frac{1}{7}M = 667652.$
on $P\ R\ \frac{2A}{7}\ (w' + w'')\frac{5}{7}c + \frac{1}{7}M = 667652.$

As the left half of the Top must be made as strong as the right half, to provide for the case of the train coming from the right, the last four compressions are the ones which we will use for the corresponding segments of the left half of the Top.

In obtaining these compressions, certain forces that acted to diminish the compressions at particular joints were neutralized as far as the top compressions are concerned. This neutralization took place through the fastenings, and therefore these forces must be taken up and provided for in getting the shearing strains on the joint pins, or other similar intermediate parts. Attention is called to this as an important point in actual construction, though immaterial in this comparison of strains.

EXTRA STRAINS.

In the preceding discussion of the strains in a truss, and the method of calculating them, we have shown that there are certain extra strains to be provided for, arising from the impossibility of keeping all the points of contact of the segments of the Top Chord in the same right line. Some of the joints have a tendency to rise, and some to fall. To keep this deflection from injuring the truss, we must make some arrangement to counteract this tendency. This is done by the introduction of panel ties. In the Fink truss the ties that support the sixteenth posts are also panel ties, and fulfil their office as such. But these only act to keep the middle, quarter, and eighth joints from rising, and to support the feet of the sixteenth posts, and thereby to keep the sixteenth joints from falling. We need, besides, panel ties to keep the sixteenth joints from rising, and to keep the middle, quarter, and eighth joints from falling. These last panel ties Fink himself does not use ; but as we consider them essential to stability, and as we deem it necessary in this comparison of the different truss combinations to make each one complete where it seems deficient, we have introduced them. To show more clearly the parts that we have added, they are represented on the drawing by dotted red lines. If the length of the segments of the Top Chord were two panels instead of one, we could dispense with these last panel ties as far as the necessity of keeping the sixteenth joints from rising is concerned ; but besides the great difficulty of making castings twenty-five feet long, they require much more iron than the two shorter segments, as we can readily see from the formula for the weight of a hollow cast-iron pillar. Calling P' the weight of the long segment, and P'' the weight of one of the short ones, we have

$$P' : 2\, P'' :: (2l)^{1.9} : 2\, (l)^{1.9}$$
$$P' : 2\, P'' :: 2^{1.9} : 2 :: 3.732 : 2$$
$$P' = 1.87 \times 2\, P''$$

In other words, the long segment would require for the same strength 87 per cent. more material than the two short ones. Moreover, if these panel ties are not on hand to resist the falling of the middle. quarter, and eighth joints, these joints will have to be upheld by the principal ties, at a cost of much more wrought-iron, or a diminution of the Factor of Safety. It is hardly probable that any method of connecting the segments of the Top at the sixteenth joints, so as to make them practically one, can be devised that will not necessitate, to secure the same strength in the two joined segments as in the single long one, at least the same amount of metal.

Total Compressions on Posts.

The measure of the deflection for which we provide is one-seventeenth of the compression of the Top. All the posts are under their maxima compressions when the head of the train is at q, except $L\,l$. This post, however, only loses, as the engine passes on to q, its portions of the excesses of the weights of the engine and of the tender over the car weight, and this loss is more than replaced by the increase in the amount of the one-seventeenth of the top compression at L, when the head of the train is at q, over what it is when the head of the train is at m. We will therefore add to the compressions when the bridge is fully loaded, one-seventeenth of the then compressions at the joints of the Top, and we will therefore have on the posts

$$\begin{aligned}
\text{on } I\,i \quad & 161989 + \tfrac{1}{17} \times 649731 = 200208 \\
\text{on } K\,k \quad & \phantom{161989 +{}} \tfrac{1}{17} \times 661198 = 38894 \\
\text{on } L\,l \quad & 22527 + \tfrac{1}{17} \times 661198 = 61421 \\
\text{on } M\,m \quad & \phantom{161989 +{}} \tfrac{1}{17} \times 661198 = 38894 \\
\text{on } N\,n \quad & 73173 + \tfrac{1}{17} \times 661198 = 112067 \\
\text{on } O\,o \quad & \phantom{161989 +{}} \tfrac{1}{17} \times 667652 = 39274 \\
\text{on } P\,p \quad & 26255 + \tfrac{1}{17} \times 667652 = 65529 \\
\text{on } Q\,q \quad & \phantom{161989 +{}} \tfrac{1}{17} \times 668132 = 39302 \\
\text{on } R\,r \quad & \phantom{161989 + \tfrac{1}{17} \times 668132} = 181715
\end{aligned}$$

Tensions on Panel Ties.

The same tensions are developed in the Panel Ties by their resistances to the upward movements of the joints to which their upper extremities are attached, as are developed in supporting the posts under the adjacent joints. Hence we have:

IRON TRUSS BRIDGES FOR RAILROADS.   59

$$\text{on } i\,K \ \tfrac{1}{34} \times 661198 \times \tfrac{1}{3}\sqrt{13} = 23372$$
$$\text{on } K\,l \ \tfrac{1}{34} \times 661198 \times \tfrac{1}{3}\sqrt{13} = 23372$$
$$\text{on } l\,M \ \tfrac{1}{34} \times 661198 \times \tfrac{1}{3}\sqrt{13} = 23372$$
$$\text{on } M\,n \ \tfrac{1}{34} \times 661198 \times \tfrac{1}{3}\sqrt{13} = 23372$$
$$\text{on } n\,O \ \tfrac{1}{34} \times 667652 \times \tfrac{1}{3}\sqrt{13} = 23601$$
$$\text{on } O\,p \ \tfrac{1}{34} \times 667652 \times \tfrac{1}{3}\sqrt{13} = 23601$$
$$\text{on } p\,Q \ \tfrac{1}{34} \times 668132 \times \tfrac{1}{3}\sqrt{13} = 23617$$
$$\text{on } Q\,r \ \tfrac{1}{34} \times 668132 \times \tfrac{1}{3}\sqrt{13} = 23617$$

Total Tension on Ties.

The shorter ties of the main system, that have the same angle of inclination as the panel ties proper, are the only ones on which any extra strains from the deflection of the Top Chord can come. The greatest total strains will come upon them from their office of supporting the posts under the sixteenth joints, as the tendency to flexure upwards at the other joints will generally be overcome by the transmitted weights. The ties between the middle and quarter posts sustain less than their maxima direct strains at the time when their total strains are the greatest, as the engine and tender are then between the quarter and the end posts. We then have on these ties

$$\text{on } k\,I \ [\tfrac{1}{2}(w'+w'') + \tfrac{1}{34}\times 661198] \times \tfrac{1}{3}\sqrt{13} = 36909.4$$
$$\text{on } k\,L \ [\tfrac{1}{2}(w'+w'') + \tfrac{1}{34}\times 661198] \times \tfrac{1}{3}\sqrt{13} = 36909.4$$
$$\text{on } m\,L \ [\tfrac{1}{2}(w'+w'') + \tfrac{1}{34}\times 661198] \times \tfrac{1}{3}\sqrt{13} = 36909.4$$
$$\text{on } m\,N \ [\tfrac{1}{2}(w'+w'') + \tfrac{1}{34}\times 661198] \times \tfrac{1}{3}\sqrt{13} = 36909.4$$
$$\text{on } o\,N \ [\tfrac{1}{2}(w'+w''+t) + \tfrac{1}{34}\times 667652] \times \tfrac{1}{3}\sqrt{13} = 38945.1$$
$$\text{on } o\,P \ [\tfrac{1}{2}(w'+w''+t) + \tfrac{1}{34}\times 667652] \times \tfrac{1}{3}\sqrt{13} = 38945.1$$
$$\text{on } q\,P \ [\tfrac{1}{2}(w'+w''+e) + \tfrac{1}{34}\times 668132] \times \tfrac{1}{3}\sqrt{13} = 39827.5$$
$$\text{on } q\,R \ [\tfrac{1}{2}(w'+w''+e) + \tfrac{1}{34}\times 668132] \times \tfrac{1}{3}\sqrt{13} = 39827.5$$

The tension on the longer ties are the same that we found before.

Bottom Horizontal Posts.

Besides the parts on which strains have been calculated, this bridge has horizontal posts between the points of support, along the line where the Bottom Chord would be, to keep these points in their proper relative places. Their dimensions depend upon the judgment of the engineer, and experience alone seems to be the guide in determining them. These posts are omitted entirely from the discussion, though they add to the weight and cost of the bridge.

Collecting together the parts that resist compression, and calculating their weights by the formula previously determined, viz.:

$$P = 0.00459961 \times W^{\frac{1}{1.19}} \times l^{\frac{1.19}{3.38}},$$

we obtain the following:

COMPRESSIONS.

NAMES.	LENGTH.	STRAIN.	FIVE TIMES THE STRAIN.	LDS. OF CAST IRON.
Top Segment $R\,Q$	12.5	668132	3340660	1665.87
" " $Q\,P$	12.5	668132	3340660	1665.87
" " $P\,O$	12.5	667652	3338260	1665.23
" " $O\,N$	12.5	667652	3338260	1665.23
" " $N\,M$	12.5	661198	3305990	1656.65
" " $M\,L$	12.5	661198	3305990	1656.65
" " $L\,K$	12.5	661198	3305990	1656.65
" " $K\,I$	12.5	661198	3305990	1656.65
Post $R\,r$	18.75	181715	908575	1803.77
" $Q\,q$	18.75	39274	196370	798.56
" $P\,p$	18.75	65529	327645	1048.49
" $O\,o$	18.75	39274	196370	798.56
" $N\,n$	18.75	112067	560335	1394.84
" $M\,m$	18.75	38894	194470	794.44
" $L\,l$	18.75	61421	307105	1013.00
" $K\,k$	18.75	38894	194470	794.44
Total				21734.90
Multiply by 2 for the other half of the truss				43469.80
Post $I\,i$	18.75	200208	1001040	1899.20
Amount of cast-iron				45369.00

Collecting together the parts that resist tension, and calculating their weights by the formula previously determined, viz. :

$$T = \frac{Wl}{18000}$$

we obtain the following :

IRON TRUSS BRIDGES FOR RAILROADS.   61

TENSIONS.

NAME.	LENGTH.	STRAIN.	FIVE TIMES THE STRAIN.	LBS. OF WROUGHT IRON.
Tie $k\,I$	22.53	36909.4	184547.0	231.04
" $k\,L$	22.53	36909.4	184547.0	231.04
" $m\,L$	22.53	36909.4	184547.0	231.04
" $m\,N$	22.53	36909.4	184547.0	231.04
" $o\,N$	22.53	38945.1	194725.5	243.78
" $o\,P$	22.53	38945.1	194725.5	243.78
" $q\,P$	22.53	39827.5	199137.5	249.31
" $q\,R$	22.53	39827.5	199137.5	249.31
" $l\,I$	31.25	44358	221790	385.05
" $l\,N$	31.25	44358	221790	385.05
" $p\,N$	31.25	44358	221790	385.05
" $p\,R$	31.25	44358	221790	385.05
" $n\,I$	53.4	140560	702300	2084.97
" $n\,R$	53.4	140560	702300	2084.97
" $i\,R$	101.74	500617	2503085	14148.36
Panel Ties $i\,K$	22.53	23372	116860	146.30
" " $K\,l$	22.53	23372	116860	146.30
" " $l\,M$	22.53	23372	116860	146.30
" " $M\,n$	22.53	23372	116860	146.30
" " $n\,O$	22.53	23601	118005	147.73
" " $O\,p$	22.53	23601	118005	147.73
" " $p\,Q$	22.53	23617	118085	147.83
" " $Q\,r$	22.53	23617	118085	147.83
Total for one half of the truss...				22945.16
Amount of wrought-iron............				45890.32

THE BOLLMAN TRUSS.

This truss, like Fink's, is properly a trussed girder, there being no Bottom Chord. It has also, like the preceding, a line of horizontal posts where the Bottom Chord would otherwise have been, to keep the points of support at their proper relative distances apart. The theory of the combination is exceedingly simple. Each point of support is directly connected with the extremities of the Top Chord by ties. In each panel there are panel ties to keep the Top Chord in line.

Assumed Weights and Dimensions.

$a\,r$ .... $= 200'$
No. of panels .... $= 16$
Panel length .... $= 12'\ 6''$
Depth of truss .... $= 18'\ 9''$
Panel weight of engine .... $= 17600 = w' + e$
    "   "   tender .... $= 16160 = w' + t$
    "   "   cars .... $= 13152 = w'$
    "   "   bridge .... $= 9375 = w$
Excess of panel weight of engine over panel weight of cars .... $= 4448 = e$
    "   "   "   tender " " " .... $= 3008 = t$

Transmission of Strains.

Each weight is transmitted to the abutments entirely independently of any other weight. The weight at m, for instance, is transmitted through the ties $m\,A$ and $m\,R$ to A and R, eleven-sixteenths of the weight going to R and five-sixteenths to A. This can at once be seen by drawing the parallelogram of forces at m. The strains on the ties are equal to the portions of the weight that travels by each, multiplied by the secant of the angle of inclination with the vertical. The compressions of the Top Chord generated by the weight at m are:

At A $\frac{11}{16} \times \frac{4}{16}\,W = \frac{44}{16}\,W$ — acting to the right.
At R $\frac{5}{16} \times \frac{11}{16}\,W = \frac{55}{16}\,W$ — acting to the left.

We see that these two compressions are equal and opposite, and that the compression of the Top Chord is the same whether we calculate it from A or from R. As we shall always assume the train as coming on from a, we will only calculate the compressions from R.

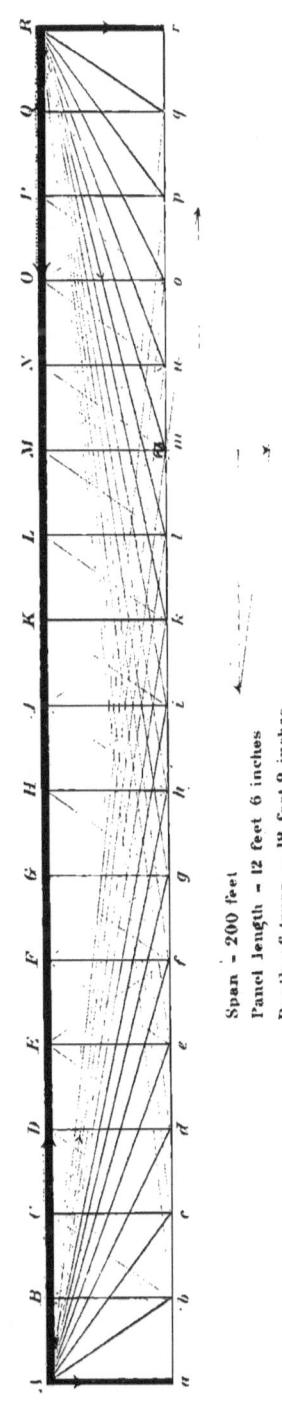

BOLLMAN

Span - 200 feet
Panel length - 12 feet 6 inches
Depth of truss - 18 feet 9 inches

From Bridge and Moving Load.

The weights from the bridge itself and from the moving load will be calculated together; for, as in Fink's truss, the counterbalancing of equal weights similarly situated does not enter the calculation in this combination.

Maxima Direct Compressions on Posts.

As we have agreed to suppose the posts of the bridge as without weight, representing the bridge weight by equal weights of 9375 lbs attached to the points of support, we will have no direct compressions on any of the posts except the end ones. These, of course, are under their maxima compressions when the bridge is covered by a train. We then have on the post Rr:

From the weight at b $\frac{1}{18}(w' + w'')$
"   " c $\frac{2}{18}(w' + w'')$
"   " d $\frac{3}{18}(w' + w'')$
"   " e $\frac{4}{18}(w' + w'')$
"   " f $\frac{5}{18}(w' + w'')$
"   " g $\frac{6}{18}(w' + w'')$
"   " h $\frac{7}{18}(w' + w'')$
"   " i $\frac{8}{18}(w' + w'')$
"   " k $\frac{9}{18}(w' + w'')$
"   " l $1\frac{2}{18}(w' + w'')$
"   " m $1\frac{4}{18}(w' + w'')$
"   " n $1\frac{3}{18}(w' + w'' + t)$
"   " o $1\frac{3}{18}(w' + w'' + t)$
"   " p $1\frac{2}{18}(w' + w'' + e)$
"   " q $1\frac{4}{18}(w' + w'' + e)$

Adding together these independent compressions, we get

On Rr $\frac{14}{18}(w' + w'') + \frac{24}{18}e + \frac{26}{18}t = 181715$.

Maxima Direct Tensions on Ties.

As no pair of ties receives any transmitted weight, it is evident that each pair of ties will be under its maximum direct tension when the engine is resting on the roadway bearer which it supports. The greatest weight that can be put on any pair of ties is therefore $w' + w'' + e$. We therefore have

on bR $\frac{1}{18}(w' + w'' + e) \times \frac{1}{2}\sqrt{509} = 33887$
on cR $\frac{2}{18}(w' + w'' + e) \times \frac{1}{2}\sqrt{703} = 63302$
on dR $\frac{3}{18}(w' + w'' + e) \times \frac{1}{2}\sqrt{685} = 88250$
on eR $\frac{4}{18}(w' + w'' + e) \times \frac{1}{2}\sqrt{585} = 108740$
on fR $\frac{5}{18}(w' + w'' + e) \times \frac{1}{2}\sqrt{493} = 124780$

on $g\,R$ $\frac{1}{12}\,(w'+w''+e)\times\frac{1}{2}\,\sqrt{409}=136384$
on $h\,R$ $\frac{1}{12}\,(w'+w''+e)\times\frac{1}{2}\,\sqrt{333}=143572$
on $i\,R$ $\frac{1}{12}\,(w'+w''+e)\times\frac{1}{2}\,\sqrt{265}=146374$
on $k\,R$ $\frac{1}{12}\,(w'+w''+e)\times\frac{1}{2}\,\sqrt{205}=144834$
on $l\,R$ $\frac{18}{12}\,(w'+w''+e)\times\frac{1}{2}\,\sqrt{153}=139026$
on $m\,R$ $\frac{14}{12}\,(w'+w''+e)\times\frac{1}{2}\,\sqrt{109}=129079$
on $n\,R$ $\frac{10}{12}\,(w'+w''+e)\times\frac{1}{2}\,\sqrt{73}=115236$
on $o\,R$ $\frac{10}{12}\,(w'+w''+e)\times\frac{1}{2}\,\sqrt{45}=98015$
on $p\,R$ $\frac{14}{12}\,(w'+w''+e)\times\frac{1}{2}\,\sqrt{25}=78629$
on $q\,R$ $\frac{18}{12}\,(w'+w''+e)\times\frac{1}{2}\,\sqrt{13}=60787$

Maxima Compressions on Top Chord.

In this truss it is evident that all of the segments of the Top Chord are at all times under the same compression. This compression will be a maximum when the bridge is covered by a train. When the head of the train is at q we have the following compressions on the Top Chord:

From weight at b $\frac{14}{4}\,(w'+w'')$
"   " c $\frac{24}{4}\,(w'+w'')$
"   " d $\frac{32}{4}\,(w'+w'')$
"   " e $\frac{44}{4}\,(w'+w'')$
"   " f $\frac{54}{4}\,(w'+w'')$
"   " g $\frac{62}{4}\,(w'+w'')$
"   " h $\frac{64}{4}\,(w'+w'')$
"   " i $\frac{64}{4}\,(w'+w'')$
"   " k $\frac{64}{4}\,(w'+w'')$
"   " l $\frac{62}{4}\,(w'+w'')$
"   " m $\frac{54}{4}\,(w'+w'')$
"   " n $\frac{44}{4}\,(w'+w''+t)$
"   " o $\frac{32}{4}\,(w'+w''+t)$
"   " p $\frac{24}{4}\,(w'+w''+e)$
"   " q $\frac{14}{4}\,(w'+w''+e)$

Adding all these independent compressions together, we obtain the following uniform maximum compression:

on $A\,R$ $\frac{440}{4}\,(w'+w'')+\frac{44}{4}c+\frac{84}{4}l=657138$.

Extra Strains.

The joints of the Top Chord are kept from rising by the panel ties. They are kept from falling by the posts, which are themselves upheld by the panel ties only, except at the ends of the truss where the ties $q\,R$ and $b\,A$ are called upon to assist. These are the only ones of the ties proper that have to be increased in section to meet this contingent strain. The tensions on the panel ties generated by holding down the joints are equal to those generated in supporting the posts that keep up the adjacent joints. The extra strains come only on the posts, on the panel ties, and on the two shortest of the ties proper. We therefore have

Total Compressions on Posts.

on $Ii \ldots \frac{1}{17} \times 657138 = 38655,$

and the same on all the others, except the end posts, which have no increase of compression.

Total Tensions on Ties.

on $qR \ldots 60787 + \frac{1}{34} \times 657138 \times \frac{1}{3} \sqrt{13} = 84016.$

On all the other ties the tensions are unchanged.

Tensions on Panel Ties.

on $Ik \ldots \frac{1}{34} \times 657138 \times \frac{1}{3} \sqrt{13} = 23220,$

and the same on all the other Panel Ties.

BOTTOM HORIZONTAL POSTS.

As in Fink's truss, the bottom horizontal posts have not entered into this calculation, but attention is called to them as members that are always used, their dimensions being decided upon by experience or the judgment of the builder. They add to the weight and cost of the bridge.

Collecting together the parts that resist compression, and calculating their weights by the formula previously determined, viz.,

$$P = 0.00459961 \times W^{\frac{1}{1.68}} \times l^{\frac{1.72}{0.91}},$$

we obtain the following:

COMPRESSIONS.

NAME.	LENGTH.	STRAIN.	FIVE TIMES THE STRAIN.	LBS. OF CAST-IRON.
Top segment $R\,Q$	12.5	657138	3285690	1651.23
" " $Q\,P$	12.5	657138	3285690	1651.23
" " $P\,O$	12.5	657138	3285690	1651.23
" " $O\,N$	12.5	657138	3285690	1651.23
" " $N\,M$	12.5	657138	3285690	1651.23
" " $M\,L$	12.5	657138	3285690	1651.23
" " $L\,K$	12.5	657138	3285690	1651.23
" " $K\,I$	12.5	657138	3285690	1651.23
Post $R\,r$	18.75	181715	908575	1803.77
" $Q\,q$	18.75	38655	193275	791.84
" $P\,p$	18.75	38655	193275	791.84
" $O\,o$	18.75	38655	193275	791.84
" $N\,n$	18.75	38655	193275	791.84
" $M\,m$	18.75	38655	193275	791.84
" $L\,l$	18.75	38655	193275	791.84
" $K\,k$	18.75	38655	193275	791.84
Total				20564.49
Multiply by 2 for the other half of the truss				41128.98
Post $I\,i$	18.75	38655	193275	791.84
Amount of cast-iron				41920.82

Collecting together the parts that resist tension, and calculating their weights by the formula previously determined, viz.,

$$T = \frac{Wl}{18000},$$

we obtain the following:

IRON TRUSS BRIDGES FOR RAILROADS. 67

TENSIONS.

NAME.	LENGTH.	STRAIN.	5 TIMES THE STRAIN.	LBS. OF WROUGHT-IRON.
Tie $b\,R$	188.44	33887	169435	1773.75
" $c\,R$	176.00	63302	316510	3094.79
" $d\,R$	163.58	88250	441250	4009.94
" $e\,R$	151.17	108740	543700	4566.09
" $f\,R$	138.77	124780	623900	4798.95
" $g\,R$	126.40	136384	681920	4788.53
" $h\,R$	114.05	143572	717860	4548.51
" $i\,R$	101.74	146374	731870	4136.80
" $k\,R$	89.49	144834	724170	3600.18
" $l\,R$	77.81	139026	695130	2985.51
" $m\,R$	65.25	129079	645395	2339.25
" $n\,R$	53.40	115236	576180	1709.33
" $o\,R$	41.93	98015	490075	1141.50
" $p\,R$	31.25	78629	393145	682.54
" $q\,R$	22.53	84016	420080	525.91
Panel Tie $r\,Q$	22.53	23229	116145	145.41
14 other Panel Ties	315.49	23229	116145	2035.67
Total for one-half of the truss				46882.66
Amount of wrought-iron				93765.32

THE JONES TRUSS.

This truss is known under different names, according to the material of which it is constructed. With all the parts of wood it is the Long truss, and this is its earliest form. With vertical iron instead of wooden ties, and some minor modifications, it is the well-known Howe truss. With the further modification of making every part of iron it becomes the Jones truss. As far as the form and theory of the combination are concerned the three trusses are the same; the changes are only those that naturally arise from the substitution of iron for wood, as the manufacture of iron increased in extent and diminished in relative cost. As we are dealing especially with iron bridges, we have called it the Jones truss.

It consists of a horizontal Top Chord of cast-iron, to resist compression; a parallel Bottom Chord of wrought-iron, to resist tension; vertical ties of wrought-iron connecting the two chords; and inclined struts of cast-iron between the top of one vertical tie and the foot of the adjacent one. It is proper to remark that Jones makes the ends of his trusses somewhat more complicated than represented in this treatise, but the object is merely to obtain a more equitable pressure on the abutment, and is rather a mechanical advantage than a theoretical necessity, and it has been considered as one of those matters of practical detail that need not here be considered. It adds, however, to the weight and cost of his bridge.

Assumed Weight and Dimensions.

$a\,r$ .... $= 200'$
Number of panels .... $= 16$
Panel length .... $= 12'\ 6''$
Depth of truss .... $= 18'\ 9''$
Panel weight of engine .... $= 17600 = w' + e$
    " " " tender .... $= 16160 = w' + t$
    " " " cars .... $= 13152 = w'$
    " " " bridge .... $=\ 9375 = w''$
$$e = 4448$$
$$t = 3008$$

Transmission of Strains.

In trusses proper, like this and the ones that follow, the principle of the counterbalancing of equal weights similarly situated is of very great importance, and is essential to determining accurately the strains on the different members. Moreover, the character of the combination is such as to necessitate the resolution of the weights along theoretical

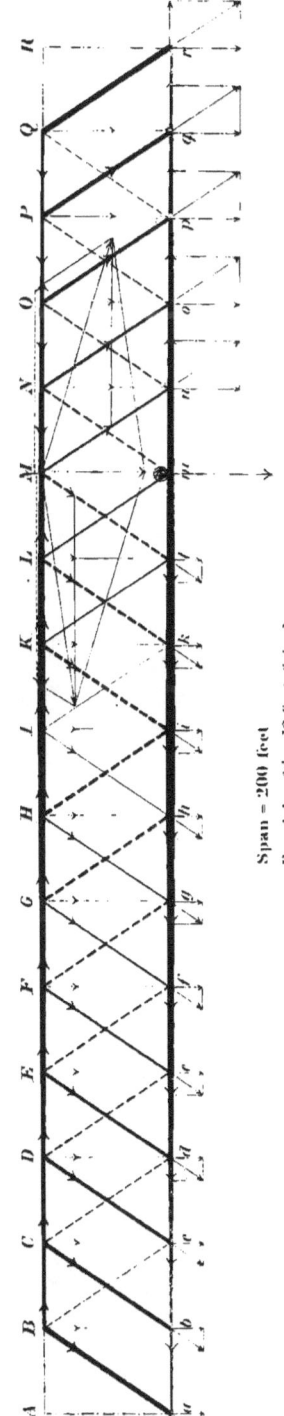

Pl.V.

JONES.

Span = 200 feet
Panel length = 12 feet 6 inches
Depth of truss = 18 feet 9 inches

lines to the points of support, as explained in the preliminary treatise on the method of calculating strains in a truss. We will illustrate the modes of ascertaining the original and deduced forces by supposing a weight at m and following its effects on the different members. It is at once carried by the vertical tie to M, and there is resolved into components along the theoretical lines Ma and Mr. First following the one along Mr we find it generating a compression at M towards Q, equal to $\frac{2}{3} \times \frac{11}{16} W = \frac{22}{48} W$. But the component along Ma at the same time generates a compression at M towards B equal to $\frac{20}{3} \times \frac{5}{16} W = \frac{100}{48} W$. The difference between these two, or $\frac{13}{48} W$, is evidently the actual effective compression at M, and it acts towards B. The component on Mr generates on Mn a compression equal to $\frac{11}{16} W \times \frac{1}{3} \sqrt{13}$. At n a tension equal to $\frac{2}{3} \times \frac{11}{16} W$, and acting towards r, is generated. Following up the strains we find—on nN a tension of $\frac{11}{16} W$; on NM, acting towards B, a compression of $\frac{2}{3} \times \frac{11}{16} W$; on No a compression equal to $\frac{11}{16} W \times \frac{1}{3} \sqrt{13}$; at o a tension, acting towards r, of $\frac{2}{3} \times \frac{11}{16} W$; on oO a tension equal to $\frac{11}{16} W$; on ON a compression acting towards B equal to $\frac{2}{3} \times \frac{11}{16} W$; on Op a compression equal to $\frac{11}{16} W \times \frac{1}{3} \sqrt{13}$; at p a tension, acting towards r, of $\frac{2}{3} \times \frac{11}{16} W$; on pP a tension equal to $\frac{11}{16} W$; on PO a compression acting towards B equal to $\frac{2}{3} \times \frac{11}{16} W$; on Pq a compression equal to $\frac{11}{16} W \times \frac{1}{3} \sqrt{13}$; at q a tension acting towards r equal to $\frac{2}{3} \times \frac{11}{16} W$; on qQ a tension equal to $\frac{11}{16} W$; on QP a compression acting towards B equal to $\frac{2}{3} \times \frac{11}{16} W$; on Qr a compression equal to $\frac{11}{16} W \times \frac{1}{3} \sqrt{13}$; at r, acting to the right, a tension equal to $\frac{2}{3} \times \frac{11}{16} W$; and on the abutment at r a compression equal to $\frac{11}{16} W$.

We find that except at M each top compression, as we go towards the abutment r, is equal to $\frac{22}{48} W$; that the bottom tensions are all $\frac{22}{48} W$; that the vertical ties have all the same tension, $\frac{11}{16} W$; and that the inclined struts have all the same compression, $\frac{11}{16} W \times \frac{1}{3} \sqrt{13}$. The top compressions and the bottom tensions are cumulative towards the middle of the truss.

Tracing up in a similar manner the effects of the component along Ma, we find at each joint of the Top Chord to the left of M a compression equal to $\frac{2}{3} \times \frac{5}{16} W = \frac{10}{48} W$, acting towards Q. At each joint of the Bottom Chord to the left of m we have a tension of $\frac{2}{3} \times \frac{5}{16} W = \frac{10}{48} W$; on each Vertical Tie a tension equal to $\frac{5}{16} W$; and on each inclined Strut a compression equal to $\frac{5}{16} W \times \frac{1}{3} \sqrt{13}$.

Adding up the top compressions in the right part of the truss, we find them increase from $\frac{22}{48} W$ at Q to $\frac{100}{48} W$ at M. In the left part we find them increase from $\frac{10}{48} W$ at B to $\frac{100}{48} W$ at L. Therefore the top compressions neutralize each other through LM. The bottom tensions on the right are $\frac{22}{48} W$ at r, and $\frac{110}{48} W$ at n; on the left they are $\frac{10}{48} W$ at a, and $\frac{110}{48} W$ at l. The bottom tensions also neutralize each other through ln. We may therefore infer that we have correctly resolved the original weight.

From Bridge Weight only.

The bridge weights are constant in position and in magnitude, and they are all equal. Hence those that are similarly situated counterbalance each other, and the inclined parts between them are unacted upon. For illustration, let us take the bridge weights at l and g. The entire weight at l goes to L and is thence transmitted undiminished through Lm and the struts parallel to it to r. Similarly the whole of the weight at g goes through Gf and the struts parallel to it to a. The top compressions which these weights generate neutralize each other through GL, and the bottom tensions through fm, no strain whatever coming on the inclined and vertical members between lL and gG.

Compressions on Struts.

on Ik .... $\frac{1}{2} w'' \times \frac{1}{3} \sqrt{13} = 4687.5 \times \frac{1}{3} \sqrt{13}$
on Kl .... $\frac{3}{2} w'' \times \frac{1}{3} \sqrt{13} = 14062.5 \times \frac{1}{3} \sqrt{13}$
on Lm .... $\frac{5}{2} w'' \times \frac{1}{3} \sqrt{13} = 23437.5 \times \frac{1}{3} \sqrt{13}$
on Mn .... $\frac{7}{2} w'' \times \frac{1}{3} \sqrt{13} = 32812.5 \times \frac{1}{3} \sqrt{13}$
on No .... $\frac{9}{2} w'' \times \frac{1}{3} \sqrt{13} = 42187.5 \times \frac{1}{3} \sqrt{13}$
on Op .... $\frac{11}{2} w'' \times \frac{1}{3} \sqrt{13} = 51562.5 \times \frac{1}{3} \sqrt{13}$
on Pq .... $\frac{13}{2} w'' \times \frac{1}{3} \sqrt{13} = 60937.5 \times \frac{1}{3} \sqrt{13}$
on Qr .... $\frac{15}{2} w'' \times \frac{1}{3} \sqrt{13} = 70312.5 \times \frac{1}{3} \sqrt{13}$

Tensions on Bottom Chord.

on rq .... $\frac{3}{2} w'' = 46875$
on qp .... $\frac{7}{2} w'' = 87500$
on po .... $\frac{13}{2} w'' = 121875$
on on .... $\frac{16}{2} w'' = 150000$
on nm .... $\frac{37}{2} w'' = 171875$
on ml .... $\frac{40}{2} w'' = 187500$
on lk .... $\frac{42}{2} w'' = 196875$
on ki .... $\frac{64}{3} w'' = 200000$

Tensions on Vertical Ties.

on Ii .... $w'' = 9375$
on Kk .... $\frac{3}{2} w'' = 14062.5$
on Ll .... $\frac{5}{2} w'' = 23437.5$
on Mm .... $\frac{7}{2} w'' = 32812.5$
on Nn .... $\frac{9}{2} w'' = 42187.5$
on Oo .... $\frac{11}{2} w'' = 51562.5$
on Pp .... $\frac{13}{2} w'' = 60937.5$
on Qq .... $\frac{15}{2} w'' = 70312.5$

Compressions on Top Chord.

on QP .... $\tfrac{1}{3}\text{fl.}\, w'' = 46875$
on PO .... $\tfrac{2}{3}\text{fl.}\, w'' = 87500$
on ON .... $\tfrac{3}{3}\text{fl.}\, w'' = 121875$
on NM .... $\tfrac{4}{3}\text{fl.}\, w'' = 150000$
on ML .... $\tfrac{5}{3}\text{fl.}\, w'' = 171875$
on LK .... $\tfrac{6}{3}\text{fl.}\, w'' = 187500$
on KI .... $\tfrac{7}{3}\text{fl.}\, w'' = 196875$

Effect of Moving Load.

In calculating the permanent strains on the bridge from its own weight we find that on certain parts, the counter-struts, we have found no strains at all. This is always the case when a bridge is uniformly loaded, and hence the necessity of actually running the train on, panel by panel. We will endeavor to ascertain what positions of the train cause the greatest strains on struts and counter-struts.

When the engine is at b, Bc bears a portion of its weight $w' + e$. As the train passes on, the weight at b becomes $w' + t$, and then w'. ⋅ No portion of the weight of the part of the train beyond b comes to the abutment a through Bc, for all that comes to c goes up cC, and down Cb, and up bB to B, and thence through Ba to a. Hence Bc s under its greatest compression when the engine is at b. The corresponding member Qp cannot receive as great a strain with the train coming from a, for it receives none at all until the engine reaches q, and then, on account of the counter-balancing of the w' of the $w' + e$ by the similar w' at b, it can only get a part of the small quantity e. Hence the value that we determine for Bc must be adopted for Qp, to provide for the case of the train coming on from r. Similarly it may be shown that we get the greatest direct strains on the members parallel to Bc when the head of the train is under their upper extremities. To illustrate, the greatest compression on Lm is found when the head of the train is at l. But if we should find a greater strain on the corresponding strut Gf, we would have to adopt that for them both. But the greatest strain on Gf, with the train coming from the left, is when the engine is at l, as then the w's at g and h and half of that at i are transmitted through Gf, together with portions of e and t. As the engine moves on, the same w's are still transmitted through Gf, but the portions of e and t are lessened. But this maximum on Gf is evidently much less than that on Lm. The same reasoning applies to all the other struts. Hence we have the following:

Maxima Direct Compressions on Counter-Struts.

Head of train at b — on Bc .... $(\frac{1}{12}w' + \frac{1}{12}e) \times \frac{1}{3}\sqrt{13} = 1100 \times \frac{1}{3}\sqrt{13}$
" " c — on Cd ... $(\frac{3}{10}w' + \frac{3}{10}e) \times \frac{1}{3}\sqrt{13} = 3300 \times \frac{1}{3}\sqrt{13}$
" " d — on De .... $(\frac{6}{10}w' + \frac{6}{10}e + \frac{1}{10}l) \times \frac{1}{3}\sqrt{13} = 6510 \times \frac{1}{3}\sqrt{13}$
" " e — on Ef .... $(\frac{10}{9}w' + \frac{10}{9}e + \frac{7}{9}l) \times \frac{1}{3}\sqrt{13} = 10732 \times \frac{1}{3}\sqrt{13}$
" " f — on Fg .... $(\frac{15}{8}w' + \frac{9}{8}e + \frac{7}{8}l) \times \frac{1}{3}\sqrt{13} = 15772 \times \frac{1}{3}\sqrt{13}$
" " g — on Gh .... $(\frac{21}{7}w' + \frac{11}{7}e + \frac{7}{7}l) \times \frac{1}{3}\sqrt{13} = 21636 \times \frac{1}{3}\sqrt{13}$
" " h — on Hi .... $(\frac{28}{6}w' + \frac{13}{6}e + \frac{1}{6}l) \times \frac{1}{3}\sqrt{13} = 28322 \times \frac{1}{3}\sqrt{13}$

Maxima Direct Compressions on Struts.

Head of train at i — on Ik .... $(\frac{1}{2}w'' + \frac{34}{10}w' + \frac{14}{6}e + \frac{11}{6}l) \times \frac{1}{3}\sqrt{13} = 40517.5 \times \frac{1}{3}\sqrt{13}$
" " k — on Kl .... $(\frac{3}{2}w'' + \frac{42}{10}w' + \frac{14}{6}e + \frac{13}{6}l) \times \frac{1}{3}\sqrt{13} = 58222.5 \times \frac{1}{3}\sqrt{13}$
" " l — on Lm .... $(\frac{5}{2}w'' + \frac{54}{10}w' + \frac{17}{6}e + \frac{14}{6}l) \times \frac{1}{3}\sqrt{13} = 76749.5 \times \frac{1}{3}\sqrt{13}$
" " m — on Mn .... $(\frac{7}{2}w'' + \frac{69}{10}w' + \frac{21}{6}e + \frac{14}{6}l) \times \frac{1}{3}\sqrt{13} = 96098.5 \times \frac{1}{3}\sqrt{13}$
" " n — on No .... $(\frac{9}{2}w'' + \frac{78}{10}w' + \frac{24}{6}e + \frac{18}{6}l) \times \frac{1}{3}\sqrt{13} = 116269.5 \times \frac{1}{3}\sqrt{13}$
" " o — on Op .... $(\frac{11}{2}w'' + \frac{94}{10}w' + \frac{24}{6}e + \frac{21}{6}l) \times \frac{1}{3}\sqrt{13} = 137262.5 \times \frac{1}{3}\sqrt{13}$
" " p — on Pq .... $(\frac{13}{2}w'' + \frac{106}{10}w' + \frac{27}{6}e + \frac{22}{6}l) \times \frac{1}{3}\sqrt{13} = 159077.5 \times \frac{1}{3}\sqrt{13}$
" " q — on Qr .... $(\frac{15}{2}w'' + \frac{126}{10}w' + \frac{29}{6}e + \frac{26}{6}l) \times \frac{1}{3}\sqrt{13} = 181714.5 \times \frac{1}{3}\sqrt{13} = 218394$

The strains on the counter-struts given above must be used for the corresponding counter-struts in the right half of the truss.

The values given above, and their fluctuations, show very clearly the office of counter-braces. They carry forward to the further abutment the proper proportions of the unbalanced weights of the moving load. We see, also, that if the moving load is uniform, the counter-braces are gradually relieved as the head of the train passes them, until, when the bridge is entirely covered, there is no strain on them at all. This fact shows how incorrect it is to attempt to calculate the strains on a bridge with the weights uniformly laid on without actually carrying them on panel by panel, as a train would do in practice.

Maxima Tensions on Bottom Chord.

These will evidently be found when the entire bridge is covered by a train. Hence,

on rq .... $\frac{1}{2}w'' + \frac{2}{3}[\frac{1}{2}w' + \frac{2}{6}e + \frac{2}{6}l] = 121143$
on qp .... $\frac{2}{3}w'' + \frac{2}{3}[\frac{2}{3}w' + \frac{14}{6}e + \frac{14}{6}l] = 221355$
on po .... $\frac{3}{2}w'' + \frac{2}{3}[\frac{3}{2}w' + \frac{13}{6}e + \frac{1}{2}l] = 306875$
on on .... $\frac{4}{2}w'' + \frac{2}{3}[\frac{4}{2}w' + \frac{12}{6}e + \frac{8}{6}l] = 377060$
on nm .... $\frac{5}{2}w'' + \frac{2}{3}[\frac{5}{2}w' + \frac{3}{6}e + \frac{9}{6}l] = 430571$
on ml .... $\frac{6}{2}w'' + \frac{2}{3}[\frac{6}{2}w' + \frac{17}{6}e + \frac{7}{6}l] = 466307$
on lk .... $\frac{7}{2}w'' + \frac{2}{3}[\frac{7}{2}w' + \frac{12}{6}e + \frac{7}{6}l] = 487400$
on ki .... $\frac{8}{2}w'' + \frac{2}{3}[\frac{8}{2}w' + \frac{7}{6}e + \frac{13}{6}l] = 493476$

IRON TRUSS BRIDGES FOR RAILROADS.   73

Maxima Direct Tensions on Vertical Ties.

Each vertical tie will evidently be under its maximum tension when the strut to which it is attached at its upper extremity is itself under its maximum compression. Besides carrying the vertical weight on the strut, the tie carries the vertical weight borne at the same time by the counter-strut. Hence,

$$\begin{aligned}
\text{on } I\,i &\quad \ldots\quad 40517.5 + \tfrac{8}{9}w' + \tfrac{8}{9}e = 49317.5 \\
\text{on } K\,k &\quad \ldots\quad 58222.5 + \tfrac{7}{9}e - \tfrac{7}{9}t = 58852.5 \\
\text{on } L\,l &\quad \ldots\quad 76749.5 + \tfrac{6}{9}e \qquad\qquad = 78417.5 \\
\text{on } M\,m &\quad \ldots\quad 96098.5 + \tfrac{5}{9}e \qquad\qquad = 97488.5 \\
\text{on } N\,n &\quad \ldots\quad 116269.5 + \tfrac{4}{9}e \qquad\qquad = 117381.5 \\
\text{on } O\,o &\quad \ldots\quad 137262.5 + \tfrac{3}{9}e \qquad\qquad = 138096.5 \\
\text{on } P\,p &\quad \ldots\quad 159077.5 + \tfrac{2}{9}e \qquad\qquad = 159633.5 \\
\text{on } Q\,q &\quad \ldots\quad 181714.5 + \tfrac{1}{9}e \qquad\qquad = 181992.5
\end{aligned}$$

Maxima Compressions on Top Chord.

The maxima compressions on the Top Chord will be found when the engine is at q, and are as follows:

$$\begin{aligned}
\text{on } I\,K &\quad \ldots\quad \tfrac{8.9}{3}w'' + \tfrac{8.9}{3}w' + \tfrac{8.9}{4.4}e + \tfrac{14.4}{4.4}t = 484533.66 \\
\text{on } K\,L &\quad \ldots\quad \tfrac{8.9}{3}w'' + \tfrac{8.9}{3}w' + \tfrac{8.4}{4.4}e + \tfrac{14.4}{4.4}t = 463440 \\
\text{on } L\,M &\quad \ldots\quad \tfrac{5.5}{3}w'' + \tfrac{5.5}{3}w' + \tfrac{5.9}{4.4}e + \tfrac{14.4}{4.4}t = 427328.33 \\
\text{on } M\,N &\quad \ldots\quad \tfrac{4.4}{3}w'' + \tfrac{4.4}{3}w' + \tfrac{4.4}{4.4}e + \tfrac{14.4}{4.4}t = 376132.66 \\
\text{on } N\,O &\quad \ldots\quad \tfrac{3.3}{3}w'' + \tfrac{3.3}{3}w' + \tfrac{7.4}{4.4}e + \tfrac{14.4}{4.4}t = 308547 \\
\text{on } O\,P &\quad \ldots\quad \tfrac{2.2}{3}w'' + \tfrac{2.2}{3}w' + \tfrac{7.4}{4.4}e + \tfrac{10.9}{4.4}t = 223746.66 \\
\text{on } P\,Q &\quad \ldots\quad \tfrac{1.4}{3}w'' + \tfrac{1.4}{3}w' + \tfrac{4.4}{4.4}e + \tfrac{4.9}{4.4}t = 120957.66
\end{aligned}$$

EXTRA STRAINS.

In this truss the joints of the top are prevented from rising by the vertical ties, and therefore extra strains on these ties must be provided for. They are prevented from falling by the struts and counter-struts, on which also extra strains may come. When the downward tendency is the greatest, which occurs when the entire bridge is covered by a train, the direct strains on the struts and counter-struts are not the maxima. We must find when the sum of the compression on the strut, and the one-thirty-fourth of the compression of the top into the secant of the angle of inclination, is a maximum; which we will find for the counter-struts to be in most cases when neither of the two strains is at its own maximum; and for the struts, when the direct strain is at its maximum.

10

Total Compressions on Counter-Struts and Struts.

Name.	Position of Engine.	Direct Strain.	Compression at Joint.		Total Strain.
B c		1100 $\times \frac{1}{3} \sqrt{13}$			1100 $\times \frac{1}{3} \sqrt{13}$
C d	at c	3300 $\times \frac{1}{3} \sqrt{13}$	C	67408.33	5282.6 $\times \frac{1}{3} \sqrt{13}$
"	at o	2466 $\times \frac{1}{3} \sqrt{13}$	C	113119	5793 $\times \frac{1}{3} \sqrt{13}$
D e	at d	6510 $\times \frac{1}{3} \sqrt{13}$	D	134386.6	10462.5 $\times \frac{1}{3} \sqrt{13}$
"	at n	4932 $\times \frac{1}{3} \sqrt{13}$	D	208626.66	11068.1 $\times \frac{1}{3} \sqrt{13}$
E f	at e	10732 $\times \frac{1}{3} \sqrt{13}$	E	198915	16582.4 $\times \frac{1}{3} \sqrt{13}$
"	at m	8220 $\times \frac{1}{3} \sqrt{13}$	E	284813	16596.9 $\times \frac{1}{3} \sqrt{13}$
F g	at f	15772 $\times \frac{1}{3} \sqrt{13}$	F	257594.66	23348.3 $\times \frac{1}{3} \sqrt{13}$
"	at g	15036 $\times \frac{1}{3} \sqrt{13}$	F	281530.66	23316.3 $\times \frac{1}{3} \sqrt{13}$
"	at h	14022 $\times \frac{1}{3} \sqrt{13}$	F	301280	22883.2 $\times \frac{1}{3} \sqrt{13}$
G h	at g	21636 $\times \frac{1}{3} \sqrt{13}$	G	308781.66	30717.8 $\times \frac{1}{3} \sqrt{13}$
"	at h	20622 $\times \frac{1}{3} \sqrt{13}$	G	335021.66	30475.6 $\times \frac{1}{3} \sqrt{13}$
H i	at h	28322 $\times \frac{1}{3} \sqrt{13}$	H	347925.66	38555.1 $\times \frac{1}{3} \sqrt{13}$
"	at i	27580 $\times \frac{1}{3} \sqrt{13}$	H	378280	38706 $\times \frac{1}{3} \sqrt{13}$
I k	at i	40517.5 $\times \frac{1}{3} \sqrt{13}$	I	364081.66	51225.8 $\times \frac{1}{3} \sqrt{13}$
"	at k	31877.5 $\times \frac{1}{3} \sqrt{13}$	I	402535	43716.8 $\times \frac{1}{3} \sqrt{13}$
K l	at k	58222.5 $\times \frac{1}{3} \sqrt{13}$	K	364140	68922.5 $\times \frac{1}{3} \sqrt{13}$
"	at l	49938.5 $\times \frac{1}{3} \sqrt{13}$	K	392868	61493.5 $\times \frac{1}{3} \sqrt{13}$
L m	at l	76749.5 $\times \frac{1}{3} \sqrt{13}$	L	343315	86847 $\times \frac{1}{3} \sqrt{13}$
"	at m	70513.5 $\times \frac{1}{3} \sqrt{13}$	L	381091.66	81745.9 $\times \frac{1}{3} \sqrt{13}$
M n	at m	96098.5 $\times \frac{1}{3} \sqrt{13}$	M	318762.66	105474 $\times \frac{1}{3} \sqrt{13}$
"	at n	90406.5 $\times \frac{1}{3} \sqrt{13}$	M	346810.33	100606.8 $\times \frac{1}{3} \sqrt{13}$
N o	at n	116269.5 $\times \frac{1}{3} \sqrt{13}$	N	270039	124211.8 $\times \frac{1}{3} \sqrt{13}$
"	at o	111121.5 $\times \frac{1}{3} \sqrt{13}$	N	292719	119731 $\times \frac{1}{3} \sqrt{13}$
O p	at o	137262.5 $\times \frac{1}{3} \sqrt{13}$	O	201766.66	143196.8 $\times \frac{1}{3} \sqrt{13}$
"	at p	132658.5 $\times \frac{1}{3} \sqrt{13}$	O	217982.66	139069.8 $\times \frac{1}{3} \sqrt{13}$
P q	at p	159077.5 $\times \frac{1}{3} \sqrt{13}$	P	112301.66	162380.5 $\times \frac{1}{3} \sqrt{13}$
"	at q	155017.5 $\times \frac{1}{3} \sqrt{13}$	P	120957.66	158575.1 $\times \frac{1}{3} \sqrt{13}$

Selecting the largest total strains, we get the following maxima total strains of compression:

Counter-struts — on Bc	....	1100	$\times \frac{1}{3} \sqrt{13} =$	1322
on Cd	....	5793	$\times \frac{1}{3} \sqrt{13} =$	6962.32
on De	....	11068.1	$\times \frac{1}{3} \sqrt{13} =$	13302.1
on Ef	....	16596.9	$\times \frac{1}{3} \sqrt{13} =$	19947
on Fg	....	23348.3	$\times \frac{1}{3} \sqrt{13} =$	28061.2
on Gh	....	30717.8	$\times \frac{1}{3} \sqrt{13} =$	36918.2
on Hi	....	38706	$\times \frac{1}{3} \sqrt{13} =$	46518.8
Struts — on Ik	....	51225.8	$\times \frac{1}{3} \sqrt{13} =$	61565.7
on Kl	....	68922.5	$\times \frac{1}{3} \sqrt{13} =$	82834.6
on Lm	....	86847	$\times \frac{1}{3} \sqrt{13} =$	104377.1
on Mn	....	105474	$\times \frac{1}{3} \sqrt{13} =$	126764
on No	....	124211.8	$\times \frac{1}{3} \sqrt{13} =$	149284
on Op	....	143196.8	$\times \frac{1}{3} \sqrt{13} =$	172101
on Pq	....	162380.5	$\times \frac{1}{3} \sqrt{13} =$	195157
on Qr	....	181714.5	$\times \frac{1}{3} \sqrt{13} =$	218394

Total Tensions on Vertical Ties.

The forces transmitted through the vertical ties tend to keep in place the joints of the Top Chord, preventing them from rising. Whenever the tendency to rise is greater than this transmitted force, the ties sustain a tension equal to the tendency to rise. If, however, this upward force, at its maximum, is less than the maximum direct strain, it is obvious that no increase of tension can come on the vertical ties. The maximum compression at I, when the entire bridge is loaded, is 483100, one-seventeenth of which is 28418, a less quantity than the maximum direct strain 40317.5. Hence Ii undergoes no extra tension. The same reasoning applies, in a stronger degree, to the other vertical ties, as their direct strains are greater than the direct strain of Ii, and the compressions at the top are less than at I. Hence no extra strains come on any of the vertical ties.

Collecting together the parts that resist compression, and calculating their weights by the formula

$$P = 0.00459961 \times W^{\frac{1}{1.49}} \times l^{\frac{1.79}{5.49}},$$

we obtain the following:

COMPRESSIONS.

NAME.	LENGTH.	STRAIN.	FIVE TIMES THE STRAIN.	LBS. OF CAST-IRON.
Top segment QP	12.5	120957.66	604788.33	671.18
" " PO	12.5	223746.66	1118733.33	930.95
" " ON	12.5	308547	1542735	1104.47
" " NM	12.5	376132.66	1880663.33	1227.21
" " ML	12.5	427328.33	2136641.66	1313.40
" " LK	12.5	463440	2317200	1371.31
" " KI	12.5	484533.66	2422668.33	1404.17
Counter-strut Qp	22.53	1322	6610	186.60
" Po	22.53	6962.32	34811.6	451.55
" On	22.53	13302.1	66510.5	637.18
" Nm	22.53	19947	99735	790.42
" Ml	22.53	28061.2	140306	947.77
" Lk	22.53	36918.2	184591	1096.66
" Ki	22.53	46518.8	232594	1240.14
Strut Ik	22.53	61565.7	307828.5	1439.49
" Kl	22.53	82834.6	414173	1685.62
" Lm	22.53	104377.1	521885.5	1906.16
" Mn	22.53	126764	633820	2113.72
" No	22.53	149284	746420	2305.81
" Op	22.53	172101	860505	2487.02
" Pq	22.53	195157	975785	2659.03
" Qr	22.53	218394	1091970	2823.00
Total for one half of the truss				30792.86
Amount of cast-iron				61585.72

Collecting together the parts that resist tension, and calculating their weights by the formula

$$T = \frac{Wl}{18000},$$

we obtain the following:

IRON TRUSS BRIDGES FOR RAILROADS. 77

TENSIONS.

NAME.	LENGTH.	STRAIN.	FIVE TIMES THE STRAIN.	LBS. OF WROUGHT-IRON.
Bottom segment rq	12.5	121143	605715	420.64
" " qp	12.5	221355	1106775	768.59
" " po	12.5	306875	1534375	1065.54
" " on	12.5	377060	1885300	1309.24
" " nm	12.5	430571	2152805	1495.04
" " ml	12.5	466307	2331535	1619.12
" " lk	12.5	487400	2437000	1692.36
" " ki	12.5	493476	2467380	1713.46
Vertical tie Kk	18.75	58852.5	294262.5	306.52
" " Ll	18.75	78417.5	392087.5	408.42
" " Mm	18.75	97488.5	487442.5	507.75
" " Nn	18.75	117381.5	586907.5	611.36
" " Oo	18.75	138096.5	690482.5	719.25
" " Pp	18.75	159633.5	798167.5	831.42
" " Qq	18.75	181992.5	909962.5	947.68
Total				14417.59
Multiply by 2 for the other half of the truss				28835.18
Vertical tie Ii	18.75	49317.5	246587.5	256.86
Amount of wrought-iron				29092.04

THE MURPHY-WHIPPLE TRUSS.

The combination in this bridge is just the reverse of that in the Jones. In this the inclined parts are ties and the vertical parts are posts. It is essentially the Pratt bridge in iron, with details from Whipple's bridge. The latter gives his ties a run of two panels instead of one as in this. We have adopted the name claimed by the builders of the bridge, deeming an investigation into its origin unnecessary to our purpose.

Assumed Weights and Dimensions.

$a\, r$ .... $= 200'$
No. of panels .... $= 16$
Panel length .... $= 12'\ 6''$
Depth of truss .... $= 18'\ 9''$
Panel weight of engine .... $= 17600 = w' + e$
    "   "   tender .... $= 16160 = w' + t$
    "   "   cars .... $= 13152 = w'$
    "   "   bridge .... $= 9375\ \ = w'''$

$$e = 4448$$
$$t = 3008$$

Transmission of Strains.

This bridge being like the one that precedes it, the strains from the weights will be obtained in a similar manner. The weights will be resolved along theoretical lines from the points of support to the ends of the Top Chord, and the principle of the counterbalancing of equal weights similarly situated will be observed.

Assume a weight at m, and trace the strains developed by it. It is at once resolved into components along $R\,m$ produced, and $A\,m$ produced. The one along $R\,m$ generates a tension at m, acting towards a, equal to $\frac{1}{3} \times \frac{13}{10} W$. The one along $A\,m$ generates a tension at m, acting towards r, equal to $\frac{10}{3} \times \frac{5}{16} W$. The resultant tension at m is therefore $\frac{12}{3} W$ acting towards r. The component along $R\,m$ produced generates a tension of $\frac{13}{10} W \times \frac{1}{3} \sqrt{13}$ on mN, nO, oP, pQ and qR. It generates a compression equal to $\frac{13}{10} W$ on Nn, Oo, Pp, Qq and Rr. It generates a compresssion equal to $\frac{1}{3} \times \frac{13}{10} W$ on NM, ON, PO, QP and RQ, the compressions accumulating from the right towards M. It generates a tension on nm, on, po and qp, equal to $\frac{1}{3} \times \frac{13}{10} W$, the tensions accumulating towards m. On rq there is no tension, as the force coming down Rr is directly neutralized by the resistance of the abutment r.

Pl. VI.

MURPHY - WHIPPLE

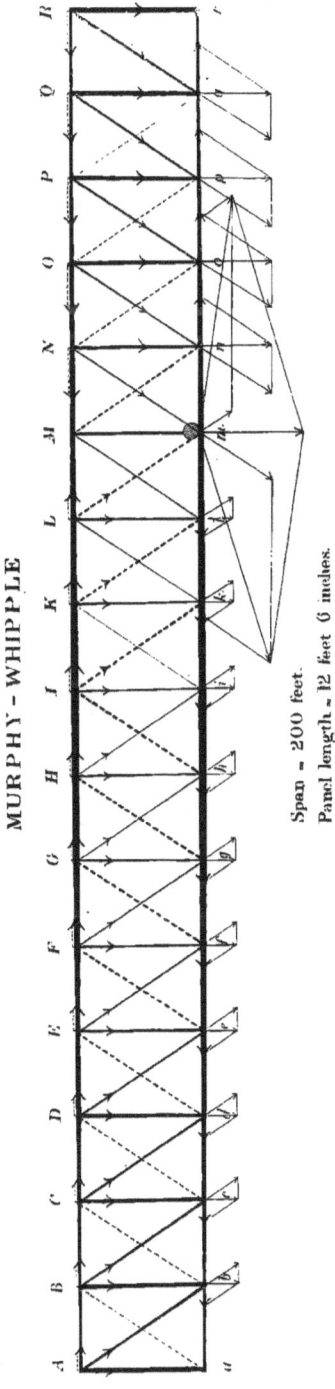

Span = 200 feet.
Panel length = 12 feet 6 inches.
Depth of truss = 18 feet 9 inches.

Similar strains are generated by the component that goes to a, the top compressions and bottom tensions neutralizing each other at M and m.

FROM BRIDGE WEIGHT ONLY.

As in Jones' truss, we find that the equal bridge weights, similarly situated, counterbalance each other. Each bridge weight is transmitted entire to its nearest abutment. Hence,

Tensions on Ties.

on iK .... $\frac{1}{2} w'' \times \frac{1}{3} \sqrt{13}$
on kL .... $\frac{3}{2} w'' \times \frac{1}{3} \sqrt{13}$
on lM .... $\frac{5}{2} w'' \times \frac{1}{3} \sqrt{13}$
on mN .... $\frac{7}{2} w'' \times \frac{1}{3} \sqrt{13}$
on nO .... $\frac{9}{2} w'' \times \frac{1}{3} \sqrt{13}$
on oP .... $\frac{11}{2} w'' \times \frac{1}{3} \sqrt{13}$
on pQ .... $\frac{13}{2} w'' \times \frac{1}{3} \sqrt{13}$
on qR .... $\frac{15}{2} w'' \times \frac{1}{3} \sqrt{13}$

Tensions on Bottom Chord.

on rq 0
on qp .... $\frac{1}{3} w''$
on po .... $\frac{2}{3} w''$
on on .... $\frac{3}{3} w''$
on nm .... $\frac{4}{3} w''$
on ml .... $\frac{5}{3} w''$
on lk .... $\frac{6}{3} w''$
on ki .... $\frac{7}{3} w''$

Compressions on Posts.

on Ii 0
on Kk .... $\frac{1}{2} w'' = 4687.5$
on Ll .... $\frac{3}{2} w'' = 14062.5$
on Mm .... $\frac{5}{2} w'' = 23437.5$
on Nn .... $\frac{7}{2} w'' = 32812.5$
on Oo .... $\frac{9}{2} w'' = 42187.5$
on Pp .... $\frac{11}{2} w'' = 51562.5$
on Qq .... $\frac{13}{2} w'' = 60937.5$
on Rr .... $\frac{15}{2} w'' = 70312.5$

Compressions on Top Chord.

on $R\,Q$ .... $\tfrac{1}{4}\,w''$
on $Q\,P$ .... $\tfrac{2\frac{1}{4}}{4}\,w''$
on $P\,O$ .... $\tfrac{4\frac{1}{2}}{4}\,w''$
on $O\,N$ .... $\tfrac{5\frac{1}{4}}{4}\,w''$
on $N\,M$ .... $\tfrac{5\frac{1}{4}}{4}\,w''$
on $M\,L$ .... $\tfrac{4\frac{1}{2}}{4}\,w''$
on $L\,K$ .... $\tfrac{2\frac{1}{4}}{4}\,w''$
on $K\,I$ .... $\tfrac{1}{4}\,w''$

EFFECT OF MOVING LOAD.

Hitherto we have found no tensions on the counter-ties, and therein they correspond with the counter-struts in the preceding bridge. As we run the train on the bridge, however, we find strains that gradually increase, and as gradually diminish, until, if the bridge is uniformly loaded, all strain is removed from the counter-ties, reappearing in the counter-ties with the opposite inclination as the train passes off the bridge.

By a similar course of reasoning to that in the discussion of the preceding bridge, we find that each one of the braces is under its maximum tension when the engine is at its foot, the brace being inclined away from the abutment from which the train comes. The braces to the left of the middle are counter-ties, those to the right are ties. Hence we have:

Maxima Direct Tensions on Counter-Ties.

on $a\,B$ .... $= 0$
on $b\,C$ .... $(\tfrac{1}{14}\,w' + \tfrac{1}{14}\,e) \times \tfrac{1}{3}\sqrt{13} = 1100 \times \tfrac{1}{3}\sqrt{13} = 1322$
on $c\,D$ .... $(\tfrac{3}{14}\,w' + \tfrac{3}{14}\,e) \times \tfrac{1}{3}\sqrt{13} = 3300 \times \tfrac{1}{3}\sqrt{13} = 3966.1$
on $d\,E$ .... $(\tfrac{6}{14}\,w' + \tfrac{6}{14}\,e + \tfrac{1}{14}\,t) \times \tfrac{1}{3}\sqrt{13} = 6510 \times \tfrac{1}{3}\sqrt{13} = 7824.1$
on $e\,F$ .... $(\tfrac{10}{14}\,w' + \tfrac{10}{14}\,e + \tfrac{3}{14}\,t) \times \tfrac{1}{3}\sqrt{13} = 10732 \times \tfrac{1}{3}\sqrt{13} = 12898.3$
on $f\,G$ .... $(\tfrac{15}{14}\,w' + \tfrac{15}{14}\,e + \tfrac{6}{14}\,t) \times \tfrac{1}{3}\sqrt{13} = 15772 \times \tfrac{1}{3}\sqrt{13} = 18955.6$
on $g\,H$ .... $(\tfrac{21}{14}\,w' + \tfrac{21}{14}\,e + \tfrac{10}{14}\,t) \times \tfrac{1}{3}\sqrt{13} = 21636 \times \tfrac{1}{3}\sqrt{13} = 26003.3$
on $h\,I$ .... $(\tfrac{28}{14}\,w' + \tfrac{28}{14}\,e + \tfrac{15}{14}\,t) \times \tfrac{1}{3}\sqrt{13} = 28322 \times \tfrac{1}{3}\sqrt{13} = 34038.8$

Maxima Direct Tensions on Ties.

on $i\,K$ .... $(\tfrac{1}{2}\,w'' + \tfrac{36}{14}\,w' + \tfrac{21}{14}\,e + \tfrac{21}{14}\,t) \times \tfrac{1}{3}\sqrt{13} = 40517.5 \times \tfrac{1}{3}\sqrt{13} = 48696$
on $k\,L$ .... $(\tfrac{3}{2}\,w'' + \tfrac{45}{14}\,w' + \tfrac{28}{14}\,e + \tfrac{28}{14}\,t) \times \tfrac{1}{3}\sqrt{13} = 58222.5 \times \tfrac{1}{3}\sqrt{13} = 69974.7$
on $l\,M$ .... $(\tfrac{6}{2}\,w'' + \tfrac{55}{14}\,w' + \tfrac{36}{14}\,e + \tfrac{36}{14}\,t) \times \tfrac{1}{3}\sqrt{13} = 76749.5 \times \tfrac{1}{3}\sqrt{13} = 92241.4$
on $m\,N$ .... $(\tfrac{7}{2}\,w'' + \tfrac{66}{14}\,w' + \tfrac{45}{14}\,e + \tfrac{45}{14}\,t) \times \tfrac{1}{3}\sqrt{13} = 96098.5 \times \tfrac{1}{3}\sqrt{13} = 115406$
on $n\,O$ .... $(\tfrac{9}{2}\,w'' + \tfrac{78}{14}\,w' + \tfrac{55}{14}\,e + \tfrac{55}{14}\,t) \times \tfrac{1}{3}\sqrt{13} = 116269.5 \times \tfrac{1}{3}\sqrt{13} = 139738.5$
on $o\,P$ .... $(\tfrac{11}{2}\,w'' + \tfrac{91}{14}\,w' + \tfrac{66}{14}\,e + \tfrac{66}{14}\,t) \times \tfrac{1}{3}\sqrt{13} = 137262.5 \times \tfrac{1}{3}\sqrt{13} = 164981$
on $p\,Q$ .... $(\tfrac{13}{2}\,w'' + \tfrac{105}{14}\,w' + \tfrac{78}{14}\,e + \tfrac{78}{14}\,t) \times \tfrac{1}{3}\sqrt{13} = 150077.5 \times \tfrac{1}{3}\sqrt{13} = 191187.4$
on $q\,R$ .... $(\tfrac{15}{2}\,w'' + \tfrac{120}{14}\,w' + \tfrac{91}{14}\,e + \tfrac{91}{14}\,t) \times \tfrac{1}{3}\sqrt{13} = 181714.5 \times \tfrac{1}{3}\sqrt{13} = 218394$

IRON TRUSS BRIDGES FOR RAILROADS. 81

Maxima Tensions on Bottom Chord.

The maxima tensions will evidently be developed in the Bottom Chord when the entire bridge is covered. Hence,

$$\text{on } r\,q \ldots\ldots = 0$$
$$\text{on } q\,p \ldots\ldots \tfrac{1.5}{3}w'' + \tfrac{1.5}{3}w' + \tfrac{5.5}{3.5}e + \tfrac{5.0}{3}\,t = 121143$$
$$\text{on } p\,o \ldots\ldots \tfrac{2.5}{3}w'' + \tfrac{2.5}{3}w' + \tfrac{9.5}{3.5}e + \tfrac{100}{3}\,t = 224303$$
$$\text{on } o\,n \ldots\ldots \tfrac{3.5}{3}w'' + \tfrac{3.5}{3}w' + \tfrac{7.8}{3.5}e + \tfrac{150}{3}\,t = 309452$$
$$\text{on } n\,m \ldots\ldots \tfrac{4.5}{3}w'' + \tfrac{3.5}{3}w' + \tfrac{7.2}{3.5}e + \tfrac{1.6.8}{3}\,t = 377032$$
$$\text{on } m\,l \ldots\ldots \tfrac{5.5}{3}w'' + \tfrac{5.5}{3}w' + \tfrac{9.0}{3.5}e + \tfrac{1.5.4}{3}\,t = 428762$$
$$\text{on } l\,k \ldots\ldots \tfrac{5.0}{3}w'' + \tfrac{5.0}{3}w' + \tfrac{5.5}{3.5}e + \tfrac{1.5.5}{3}\,t = 464873$$
$$\text{on } k\,i \ldots\ldots \tfrac{5.0}{3}w'' + \tfrac{5.0}{3}w' + \tfrac{4.5}{3.5}e + \tfrac{1.3.9}{3}\,t = 485967$$

Maxima Direct Compressions on Posts.

The maxima direct compressions on the posts will evidently be found when the braces attached to their upper extremities are under their maxima tensions. Hence,

$$\text{on } I\,i \ldots\ldots 28322 \quad -\text{weight on } h\,I$$
$$\text{on } K\,k \ldots\ldots 40517.5 - \quad " \quad i\,K$$
$$\text{on } L\,l \ldots\ldots 58222.5 - \quad " \quad k\,L$$
$$\text{on } M\,m \ldots\ldots 76749.5 - \quad " \quad l\,M$$
$$\text{on } N\,n \ldots\ldots 96098.5 - \quad " \quad m\,N$$
$$\text{on } O\,o \ldots\ldots 116269.5 - \quad " \quad n\,O$$
$$\text{on } P\,p \ldots\ldots 137262.5 - \quad " \quad o\,P$$
$$\text{on } Q\,q \ldots\ldots 159077.5 - \quad " \quad p\,Q$$
$$\text{on } R\,r \ldots\ldots 181714.5 - \quad " \quad q\,R$$

Maxima Compressions on Top Chord.

These will evidently be found when the head of the train is at q. They are

$$\text{on } I\,K \ldots\ldots \tfrac{5.5}{3}w'' + \tfrac{5.5}{3}w' + \tfrac{5.4}{3.5}e + \tfrac{1.3.8}{3}\,t = 493476$$
$$\text{on } K\,L \ldots\ldots \tfrac{5.0}{3}w'' + \tfrac{5.0}{3}w' + \tfrac{9.0}{3.5}e + \tfrac{1.4.0}{3}\,t = 487400$$
$$\text{on } L\,M \ldots\ldots \tfrac{5.0}{3}w'' + \tfrac{5.0}{3}w' + \tfrac{9.4}{3.5}e + \tfrac{1.4.5}{3}\,t = 466307$$
$$\text{on } M\,N \ldots\ldots \tfrac{4.5}{3}w'' + \tfrac{5.2}{3}w' + \tfrac{7.2}{3.5}e + \tfrac{1.4.9}{3}\,t = 430571$$
$$\text{on } N\,O \ldots\ldots \tfrac{3.5}{3}w'' + \tfrac{4.5}{3}w' + \tfrac{7.4}{3.5}e + \tfrac{1.4.4}{3}\,t = 377060$$
$$\text{on } O\,P \ldots\ldots \tfrac{3.0}{3}w'' + \tfrac{3.0}{3}w' + \tfrac{9.4}{3.5}e + \tfrac{1.4.0}{3}\,t = 306875$$
$$\text{on } P\,Q \ldots\ldots \tfrac{2.5}{3}w'' + \tfrac{2.5}{3}w' + \tfrac{5.4}{3.5}e + \tfrac{1.0.0}{3}\,t = 221355$$
$$\text{on } Q\,R \ldots\ldots \tfrac{1.5}{3}w'' + \tfrac{1.5}{3}w' + \tfrac{5.5}{3.5}e + \tfrac{5.0}{3}\,t = 121143$$

Extra Strains.

In this bridge the joints of the Top Chord are prevented from rising by the braces and the transmitted weights; and from falling by the posts.

Total Tensions on Counter-Ties and Ties.

There are no extra tensions on the braces in this truss, as the transmitted forces acting vertically downward at each joint when any brace is under its maximum direct tension are in all cases greater than one-thirty-fourth the then compression of the Top— as an examination of the Top compressions will readily show.

Total Compressions on Posts.

In addition to their direct weights, the posts must be strong enough to keep the joints of the Top Chord from deflecting when they themselves are under their maxima direct compressions. Hence,

Engine at i — on Ii .... $27580 + \frac{1}{17} \times 391093.33$ [on IK] $= 50585.5$
" at i — on Kk .... $40517.5 + \frac{1}{17} \times 364081.66$ [on KL] $= 61934.1$
" at k — on Ll .... $58222.5 + \frac{1}{17} \times 364140$ [on LM] $= 79642.5$
" at l — on Mm .... $76749.5 + \frac{1}{17} \times 349581.66$ [on MN] $= 97313.1$
" at m — on Nn .... $96098.5 + \frac{1}{17} \times 318762.66$ [on NO] $= 114849.2$
" at n — on Oo .... $116269.5 + \frac{1}{17} \times 270039$ [on OP] $= 132154.2$
" at o — on Pp .... $137262.5 + \frac{1}{17} \times 201766.66$ [on PQ] $= 149131.1$
" at p — on Qq .... $159077.5 + \frac{1}{17} \times 112301.66$ [on QR] $= 165683.5$
" at q — on Rr .... $= 181714.5$

The total compression on Ii is greater when the engine is at i than when it is at h.

Collecting together the parts that resist compression, and calculating their weights by the formula

$$v = 0.00459961 \times W^{\frac{1}{1.19}} \times l^{\frac{1.19}{1.19}},$$

we obtain the following:

IRON TRUSS BRIDGES FOR RAILROADS. 83

COMPRESSIONS.

NAME.	LENGTH.	STRAIN.	FIVE TIMES THE STRAIN.	LBS. OF CAST-IRON.
Top Segment R Q	12.5	121143	605715	671.73
" " Q P	12.5	221355	1106775	925.64
" " P O	12.5	306875	1534375	1101.30
" " O N	12.5	377060	1885300	1228.81
" " N M	12.5	430571	2152855	1318.69
" " M L	12.5	466307	2331535	1375.82
" " L K	12.5	487400	2437000	1408.58
" " K I	12.5	493476	2467380	1417.89
Post K k	18.75	61934.1	309670 5	1017.78
" L l	18.75	79642.5	398212.5	1163.12
" M m	18.75	97313.1	486565.5	1293.02
" N n	18.75	114849.2	574246	1411.43
" O o	18.75	132154.2	660771	1522.69
" P p	18.75	149131.1	745655.5	1623.70
" Q q	18.75	165683.5	828417.5	1717.30
" R r	18.75	181714.5	908572.5	1803.77
Total				21002.26
Multiply by 2 for the other half of the truss				42004.52
Post I i	18.75	50585.5	252927.5	913.64
Amount of cast-iron				42918.16

Collecting together the parts that resist tension, and calculating their weights by the formula

$$T = \frac{Wl}{18000},$$

we obtain the following:

Tensions.

NAME.	LENGTH.	STRAIN.	FIVE TIMES THE STRAIN.	LBS. OF WROUGHT-IRON.
Bottom segment $r\,q$	12.5	0	0	0
" " $q\,p$	12.5	121143	605715	420.64
" " $p\,o$	12.5	224303	1121515	778.83
" " $o\,n$	12.5	309452	1547260	1074.52
" " $n\,m$	12.5	377632	1888160	1311.22
" " $m\,l$	12.5	428762	2143810	1488.76
" " $l\,k$	12.5	464873	2324365	1614.14
" " $k\,i$	12.5	485967	2429835	1687.38
Counter-tie $r\,Q$	22.53	0	0	0
" $q\,P$	22.53	1322	6610	8.28
" $p\,O$	22.53	3966.1	19830.5	24.83
" $o\,N$	22.53	7824.1	39120.5	48.98
" $n\,M$	22.53	12898.3	64496.5	80.74
" $m\,L$	22.53	18955.6	94778	118.66
" $l\,K$	22.53	26003.3	130016.5	162.77
" $k\,I$	22.53	34038.8	170194	213.07
Tie $i\,K$	22.53	48696	243480	304.82
" $k\,L$	22.53	69974.7	349873.5	438.02
" $l\,M$	22.53	92241.4	461207	577.40
" $m\,N$	22.53	115496	577480	722.96
" $n\,O$	22.53	139738.5	698692.5	874.71
" $o\,P$	22.53	164931	824655	1032.41
" $p\,Q$	22.53	191187.4	955937	1196.76
" $q\,R$	22.53	218394	1091970	1367.06
Total for one half of the truss				15546.96
Amount of wrought-iron				31093.92

POST

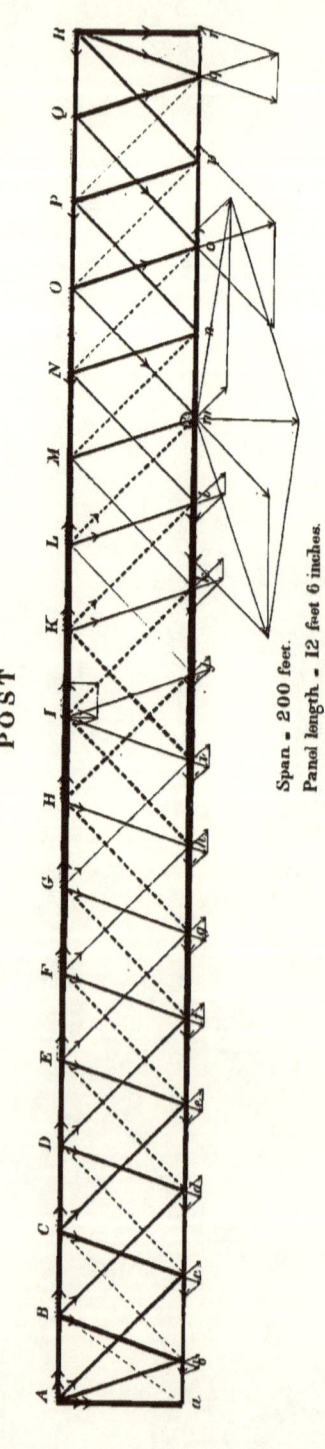

Span = 200 feet.
Panel length = 12 feet 6 inches.
Depth of truss = 18 feet 9 inches.

THE POST TRUSS.

In this truss the posts are inclined, with the run of half a panel ; the ties have a run of a panel and a half, and the counter-ties the same. Owing to the inclination of the posts, the ties run from the foot of one post to the top of the second post from it ; while the counter-ties run from the foot of one post to the top of the adjacent one. The Top Chord has a joint in the middle, but the Bottom Chord has not, having joints on either side of the middle, and half panels at each end.

Assumed Weights and Measures.

As this bridge is assumed to weigh the same as the others, we must obviously assume half bridge weights on the end joints of the Bottom Chord, otherwise the bridge will be heavier than its assumed weight. We must, also, assume half weights for the moving load when it is at either of the end joints, as the abutments, which are nearer the end joints than in the other bridges, bear a greater proportion of the train weight, the weight of the entire train being the same as in the bridges that precede. In other words, we must assume, for the sixteen points of support in this bridge, the same load as for the fifteen of those that have already been examined.

$$a\,r \ldots\ldots\ldots\ldots\ldots\ldots\ldots\ldots\ldots\ldots\ldots\ldots\ldots\ldots = 200'$$
$$\text{No. of panels} \ldots\ldots\ldots\ldots\ldots\ldots\ldots\ldots\ldots\ldots\ldots = 16$$
$$\text{Panel length} \ldots\ldots\ldots\ldots\ldots\ldots\ldots\ldots\ldots\ldots\ldots = 12'\,6''$$
$$\text{Depth of truss} \ldots\ldots\ldots\ldots\ldots\ldots\ldots\ldots\ldots\ldots\ldots = 18'\,9''$$
$$\text{Panel weight of engine} \ldots\ldots\ldots\ldots\ldots\ldots\ldots = 17600 = w' + e$$
$$\text{\hphantom{Panel weight of} tender} \ldots\ldots\ldots\ldots\ldots\ldots\ldots = 16160 = w' + t$$
$$\text{\hphantom{Panel weight of} cars} \ldots\ldots\ldots\ldots\ldots\ldots\ldots\ldots = 13152 = w'$$
$$\text{\hphantom{Panel weight of} bridge} \ldots\ldots\ldots\ldots\ldots\ldots\ldots\ldots = 9375 = w''$$
$$e = 4448$$
$$t = 3008$$

Transmission of Strains.

Equal weights similarly situated counterbalance, and each is transmitted undivided to its nearest abutment.

Assume an unbalanced weight at m. It is at once resolved into components along the prolongations of $R\,m$ and $A\,m$. The first component will be again resolved along the Bottom Chord and the tie $m\,O$. The second component, along the Bottom Chord and the counter-tie mL. The two secondary components along the Bottom Chord will give a single one equal to their difference, acting in the direction of the greater. The fact of two opposite tensions being generated at m will only affect the shearing force on the pin, or other part connecting the tie and counter-tie. These forces, in other respects combined, must be calculated separately for the shearing strains at the joints.

The component transmitted to O generates a compression of the Top Chord at O equal to $\frac{4}{3} \times \frac{23}{33} W$, the same at Q, and $\frac{1}{3} \times \frac{23}{33} W$ at R. It generates a compression on Oo equal to $\frac{23}{33} W \times \frac{1}{3}\sqrt{10}$, the same on Qq, and $\frac{23}{33} W$ on Rr. It generates a tension on the Bottom Chord at o equal to $\frac{4}{3} \times \frac{23}{33} W$, and at q equal to $\frac{3}{3} \times \frac{23}{33} W$. The tension at m equals $\frac{20}{33} \times \frac{9}{33} W - \frac{1}{3} \times \frac{23}{33} W = \frac{180}{99} W - \frac{138}{99} W = \frac{42}{99} W$.

Similar resolutions of the other component may be made. When the transmitted force reaches I it is divided between Ii and Ij, and thence is transmitted to A by halves. Otherwise the two components act exactly alike.

From Bridge Weight only.

These are all balanced weights, and go to their nearest abutments. The weights at b and q are $\frac{1}{2} w''$.

Tensions on Ties.

on jL .... $w'' \times \sqrt{2}$
on kM .... $w'' \times \sqrt{2}$
on lN .... $2 w'' \times \sqrt{2}$
on mO .... $2 w'' \times \sqrt{2}$
on nP .... $3 w'' \times \sqrt{2}$
on oQ .... $3 w'' \times \sqrt{2}$
on pR .... $4 w'' \times \sqrt{2}$
on qR .... $\frac{1}{2} w'' \times \frac{1}{2} \sqrt{10}$

Tensions on Bottom Chord.

on rq 0
on qp .... $1\frac{3}{6} w''$
on po .... $4\frac{1}{9} w''$
on on .... $5\frac{5}{6} w''$
on nm .... $8\frac{7}{8} w''$
on ml .... $10\frac{1}{6} w''$
on lk .... $11\frac{2}{3} w''$
on kj .... $12\frac{1}{6} w''$
on ji .... $12\frac{1}{6} w''$

IRON TRUSS BRIDGES FOR RAILROADS.   87

Compressions on Posts.

on Ij 0
on Kk 0
on Ll .... $w'' \times \frac{1}{3} \sqrt{10}$
on Mm .... $w'' \times \frac{1}{3} \sqrt{10}$
on Nn .... $2\ w'' \times \frac{1}{3} \sqrt{10}$
on Oo .... $2\ w'' \times \frac{1}{3} \sqrt{10}$
on Pp .... $3\ w'' \times \frac{1}{3} \sqrt{10}$
on Qq .... $3\ w'' \times \frac{1}{3} \sqrt{10}$
on Rr .... $\frac{19}{2} w''$

Compressions on Top Chord.

on RQ .... $\frac{31}{6} w'' = 48437.5$
on QP .... $\frac{55}{6} w'' = 85937.5$
on PO .... $\frac{79}{6} w'' = 123437.5$
on ON .... $\frac{95}{6} w'' = 148437.5$
on NM .... $\frac{111}{6} w'' = 173437.5$
on ML .... $\frac{119}{6} w'' = 185937.5$
on LK .... $\frac{127}{6} w'' = 198437.5$
on KI .... $\frac{127}{6} w'' = 198437.5$

EFFECT OF MOVING LOAD.

The effect of running a train on this bridge is the same as in the preceding ones. The counter-ties, on which we have found no strain from the bridge weight, are called into action to distribute the unbalanced moving load to the further abutment, the strain on them lessening after the middle is passed until, when the load is uniform, there is again no strain on them at all when the entire bridge is covered. As the load passes off, strains are developed in the counter-ties of the opposite inclination, which strains also reduce to 0 when the bridge becomes unloaded.

Running the train on from a, we find that each Counter-tie and Tie of the system whose lower extremities are nearer a than their upper, will be under its maximum direct tension when the engine reaches its lower extremity. Also, each Post will be under its maximum direct compression when the tie attached to its upper extremity is under its maximum tension. Ij and Kk have their greatest compressions when hI and iK have their greatest tensions.

The Top and Bottom Chords evidently are under their greatest strains when the entire bridge is covered by a train.

Maxima Direct Tensions on Counter-Ties.

$$
\begin{aligned}
\text{on } a\,B &\ldots\ 0 \\
\text{on } b\,C &\ldots\ (\tfrac{1}{14} w' + \tfrac{1}{14} e) \times \sqrt{2} = 275 \times \sqrt{2} = 388.9 \\
\text{on } c\,D &\ldots\ (\tfrac{7}{14} w' + \tfrac{7}{14} e) \times \sqrt{2} = 1925 \times \sqrt{2} = 2722.4 \\
\text{on } d\,E &\ldots\ (\tfrac{17}{14} w' + \tfrac{17}{14} e + \tfrac{1}{14} t) \times \sqrt{2} = 4652.5 \times \sqrt{2} = 6579.6 \\
\text{on } e\,F &\ldots\ (\tfrac{31}{14} w' + \tfrac{31}{14} e + \tfrac{7}{14} t) \times \sqrt{2} = 8367.5 \times \sqrt{2} = 11833.4 \\
\text{on } f\,G &\ldots\ (\tfrac{49}{14} w' + \tfrac{49}{14} e + \tfrac{14}{14} t) \times \sqrt{2} = 13045.5 \times \sqrt{2} = 18449.1 \\
\text{on } g\,H &\ldots\ (\tfrac{71}{14} w' + \tfrac{49}{14} e + \tfrac{22}{14} t) \times \sqrt{2} = 18498.5 \times \sqrt{2} = 26160.84 \\
\text{on } h\,I &\ldots\ (\tfrac{97}{14} w' + \tfrac{49}{14} e + \tfrac{32}{14} t) \times \sqrt{2} = 24773.5 \times \sqrt{2} = 35035.02 \\
\text{on } i\,K &\ldots\ (\tfrac{127}{14} w' + \tfrac{49}{14} e + \tfrac{44}{14} t) \times \sqrt{2} = 20060.25 \times \sqrt{2} = 28369.13
\end{aligned}
$$

Maxima Direct Tensions on Ties.

$$
\begin{aligned}
\text{on } j\,L &\ldots\ (w'' + \tfrac{161}{14} w' + \tfrac{61}{14} e + \tfrac{44}{14} t) \times \sqrt{2} = 38069.75 \times \sqrt{2} = 53838.77 \\
\text{on } k\,M &\ldots\ (w'' + \tfrac{133}{14} w' + \tfrac{49}{14} e + \tfrac{44}{14} t) \times \sqrt{2} = 33686.25 \times \sqrt{2} = 47639.55 \\
\text{on } l\,N &\ldots\ (2 w'' + \tfrac{197}{14} w' + \tfrac{73}{14} e + \tfrac{61}{14} t) \times \sqrt{2} = 56015.75 \times \sqrt{2} = 79218.24 \\
\text{on } m\,O &\ldots\ (2 w'' + \tfrac{171}{14} w' + \tfrac{49}{14} e + \tfrac{44}{14} t) \times \sqrt{2} = 53222.25 \times \sqrt{2} = 75267.62 \\
\text{on } n\,P &\ldots\ (3 w'' + \tfrac{191}{14} w' + \tfrac{49}{14} e + \tfrac{44}{14} t) \times \sqrt{2} = 74776.75 \times \sqrt{2} = 105750.3 \\
\text{on } o\,Q &\ldots\ (3 w'' + \tfrac{171}{14} w' + \tfrac{49}{14} e + \tfrac{44}{14} t) \times \sqrt{2} = 74215.25 \times \sqrt{2} = 104956.24 \\
\text{on } p\,R &\ldots\ (4 w'' + \tfrac{113}{14} w' + \tfrac{61}{14} e + \tfrac{44}{14} t) \times \sqrt{2} = 90591.75 \times \sqrt{2} = 136601.4 \\
\text{on } q\,R &\ldots\ (\tfrac{7}{14} w'' + \tfrac{148}{14} w' + \tfrac{31}{14} e + \tfrac{44}{14} t) \times \tfrac{1}{2}\sqrt{10} = 84337 \times \tfrac{1}{2}\sqrt{10} = 88694.5
\end{aligned}
$$

Maxima Tensions on Bottom Chord.

$$
\begin{aligned}
\text{on } r\,q &\ldots\ 0 \\
\text{on } q\,p &\ldots\ \tfrac{14}{14} w'' + \tfrac{14}{14} w' + \tfrac{31}{14} e + \tfrac{41}{14} t = 51149.66 \\
\text{on } p\,o &\ldots\ \tfrac{44}{14} w'' + \tfrac{44}{14} w' + \tfrac{91}{14} e + \tfrac{141}{14} t = 170484.86 \\
\text{on } o\,n &\ldots\ \tfrac{54}{14} w'' + \tfrac{54}{14} w' + \tfrac{91}{14} e + \tfrac{241}{14} t = 254827.83 \\
\text{on } n\,m &\ldots\ \tfrac{87}{14} w'' + \tfrac{87}{14} w' + \tfrac{77}{14} e + \tfrac{241}{14} t = 338481.16 \\
\text{on } m\,l &\ldots\ \tfrac{121}{14} w'' + \tfrac{121}{14} w' + \tfrac{79}{14} e + \tfrac{241}{14} t = 389968.83 \\
\text{on } l\,k &\ldots\ \tfrac{118}{14} w'' + \tfrac{118}{14} w' + \tfrac{63}{14} e + \tfrac{241}{14} t = 441606.5 \\
\text{on } k\,j &\ldots\ \tfrac{121}{14} w'' + \tfrac{121}{14} w' + \tfrac{58}{14} e + \tfrac{192}{14} t = 462905.16 \\
\text{on } j\,i &\ldots\ \tfrac{121}{14} w'' + \tfrac{121}{14} w' + \tfrac{59}{14} e + \tfrac{144}{14} t = 484894
\end{aligned}
$$

Maxima Direct Compressions on Posts.

$$
\begin{aligned}
\text{on } I\,j &\ldots\ (\tfrac{1}{14} w' + \tfrac{22}{14} e + \tfrac{44}{14} t) \times \tfrac{1}{2}\sqrt{10} = 12386.75 \times \tfrac{1}{2}\sqrt{10} \\
\text{on } K\,k &\ldots\ (\tfrac{133}{14} w' + \tfrac{49}{14} e + \tfrac{44}{14} t) \times \tfrac{1}{2}\sqrt{10} = 20060.25 \times \tfrac{1}{2}\sqrt{10} \\
\text{on } L\,l &\ldots\ (w'' + \tfrac{161}{14} w' + \tfrac{61}{14} e + \tfrac{44}{14} t) \times \tfrac{1}{2}\sqrt{10} = 38069.75 \times \tfrac{1}{2}\sqrt{10} \\
\text{on } M\,m &\ldots\ (w'' + \tfrac{133}{14} w' + \tfrac{49}{14} e + \tfrac{44}{14} t) \times \tfrac{1}{2}\sqrt{10} = 33686.25 \times \tfrac{1}{2}\sqrt{10} \\
\text{on } N\,n &\ldots\ (2 w'' + \tfrac{197}{14} w' + \tfrac{73}{14} e + \tfrac{44}{14} t) \times \tfrac{1}{2}\sqrt{10} = 56015.75 \times \tfrac{1}{2}\sqrt{10} \\
\text{on } O\,o &\ldots\ (2 w'' + \tfrac{171}{14} w' + \tfrac{49}{14} e + \tfrac{44}{14} t) \times \tfrac{1}{2}\sqrt{10} = 53222.25 \times \tfrac{1}{2}\sqrt{10} \\
\text{on } P\,p &\ldots\ (3 w'' + \tfrac{191}{14} w' + \tfrac{49}{14} e + \tfrac{44}{14} t) \times \tfrac{1}{2}\sqrt{10} = 74776.75 \times \tfrac{1}{2}\sqrt{10} \\
\text{on } Q\,q &\ldots\ (3 w'' + \tfrac{171}{14} w' + \tfrac{61}{14} e + \tfrac{44}{14} t) \times \tfrac{1}{2}\sqrt{10} = 74215.25 \times \tfrac{1}{2}\sqrt{10} \\
\text{on } R\,r &\ldots\ \tfrac{14}{14} w'' + \tfrac{14}{14} w' + \tfrac{44}{14} e + \tfrac{194}{14} t \phantom{\times \tfrac{1}{2}\sqrt{10}} = 180026
\end{aligned}
$$

Maxima Compressions on Top Chord.

These will be found when the engine is at q. They are

on IK .... $\frac{131}{2} w'' + \frac{131}{2} w' + \frac{133}{2} e + \frac{22.8}{2.2} t = 487046$

on KL .... $\frac{131}{2} w'' + \frac{131}{2} w' + \frac{144}{2} e + \frac{22.2}{2.2} t = 488123$

on LM .... $\frac{142}{2} w'' + \frac{142}{2} w' + \frac{164}{2} e + \frac{276}{2.2} t = 459063.33$

on MN .... $\frac{111}{2} w'' + \frac{111}{2} w' + \frac{175}{2} e + \frac{300}{22} t = 445828.66$

on NO .... $\frac{90}{2} w'' + \frac{90}{2} w' + \frac{168}{2} e + \frac{316}{22} t = 370769.33$

on OP .... $\frac{70}{2} w'' + \frac{70}{2} w' + \frac{203}{2} e + \frac{312}{22} t = 311021.66$

on PQ .... $\frac{51}{2} w'' + \frac{51}{2} w' + \frac{294}{2} e + \frac{210}{22} t = 217826.66$

on QR .... $\frac{31}{2} w'' + \frac{31}{2} w' + \frac{203}{2} e + \frac{192}{22} t = 124334.66$

Extra Strains.

It is impossible to keep the joints of the Top Chord from tending to deflect from the line of the chord, and we must provide for the extra strains which any tendency to flexure would bring on the parts of the truss.

The joints of the Top Chord are prevented from rising above the line of the chord by the tie and counter-tie at each joint, and by the transmitted weights at the joint. They are prevented from falling below the line of the chord by the posts, on which they necessarily bring extra strains.

Total Tensions on Counter-ties and Ties.

If any joint has a tendency to rise it will tend to bring on the tie and counter-tie a strain equal to its upward force into the secant of the angle of the brace, which is $\sqrt{2}$. But the transmitted force at the joint acts to hold it down. The greatest upward tendency at any joint is at K, where, when the entire bridge is covered, it is 487046, one-thirty-fourth of which is 14324.9. But this is less than the maximum tension already provided for on all the ties, and on four of the counter-ties. No extra strains can come on the other counter-ties, because the transmitted weights at the joints to which they are attached are in all cases greater than one-thirty-fourth the top compression at the joint. Hence there are no extra strains on ties or on counter-ties in this truss.

Total Compressions on Posts.

The posts must sustain the joints of the top from falling. The greatest total compressions on them will evidently be when they are under their maxima direct compressions. We must then add to the compressions previously determined, one-seventeenth the then compression of the Top into the secant of the angle of inclination with the vertical.

Determining then the compressions on the different segments of the Top at the time when the posts under them are under maxima direct compressions, we have as follows:

$$\begin{aligned}
\text{on } Ij &\ldots (12386.75 + \tfrac{1}{17} \times 371162.83) \times \tfrac{1}{3}\sqrt{10} = 34219.85 \times \tfrac{1}{3}\sqrt{10} = 36070.9\\
\text{on } Kk &\ldots (20060.25 + \tfrac{1}{17} \times 344416.5) \times \tfrac{1}{3}\sqrt{10} = 40320.05 \times \tfrac{1}{3}\sqrt{10} = 42501.1\\
\text{on } Ll &\ldots (38069.75 + \tfrac{1}{17} \times 339228.83) \times \tfrac{1}{3}\sqrt{10} = 58024.4 \times \tfrac{1}{3}\sqrt{10} = 61163.2\\
\text{on } Mm &\ldots (33686.25 + \tfrac{1}{17} \times 350800.83) \times \tfrac{1}{3}\sqrt{10} = 54321.6 \times \tfrac{1}{3}\sqrt{10} = 57260\\
\text{on } Nn &\ldots (56015.75 + \tfrac{1}{17} \times 297874.16) \times \tfrac{1}{3}\sqrt{10} = 73537.75 \times \tfrac{1}{3}\sqrt{10} = 77515.57\\
\text{on } Oo &\ldots (53222.25 + \tfrac{1}{17} \times 260239.16) \times \tfrac{1}{3}\sqrt{10} = 68530.45 \times \tfrac{1}{3}\sqrt{10} = 72237.4\\
\text{on } Pp &\ldots (74776.75 + \tfrac{1}{17} \times 187645.5) \times \tfrac{1}{3}\sqrt{10} = 85814.75 \times \tfrac{1}{3}\sqrt{10} = 90456.66\\
\text{on } Qq &\ldots (74215.25 + \tfrac{1}{17} \times 122687.16) \times \tfrac{1}{3}\sqrt{10} = 81432.15 \times \tfrac{1}{3}\sqrt{10} = 85837\\
\text{on } Rr &\ldots \phantom{(00000.00 + \tfrac{1}{17} \times 000000.00)\times \tfrac{1}{3}\sqrt{10} = }180026.
\end{aligned}$$

We have found no strains on aB, rQ, ab and qr. The former are chiefly used for bringing a strain on the posts Bb or Qq when the bridge is unloaded, which strain disappears when the moving load covers the bridge. They keep a constant pressure on b and q, so that when the engine comes upon the bridge it finds all the parts in close connection, ready to receive and transmit the strains that belong to them, and any shock or depression from the sudden application of the weight of the engine is avoided.

The segments ab and qr are necessary to resist the compression generated by the strain put upon aB and rQ, and also to connect b and q with the abutments and prevent longitudinal motion in the bottom of the truss from the horizontal pull made by the friction of the engine on the track.

In this truss there is a little less maximum strain on Rr than in the others. This is due to the necessarily smaller proportion of e borne by Rr in this combination, though it sustains a slightly larger proportion of t. More of e comes directly on the abutment. The absence of a joint at the middle of the Bottom Chord slightly lessens the total strains.

Collecting together the parts that resist compression, and calculating their weights by the formula

$$P = 0.00459961 \times W^{\frac{1}{1.18}} \times l^{\frac{1.18}{1.18}}$$

we obtain the following:

COMPRESSIONS.

NAME.	LENGTH.	STRAIN.	FIVE TIMES THE STRAIN.	LBS. OF CAST-IRON.
Top segment R Q	12.5	124334.66	621673.3	681.11
" " Q P	12.5	217826.66	1089133.3	915.52
" " P O	12.5	311021.66	1555108.3	1109.45
" " O N	12.5	370769.33	1853846.7	1217.87
" " N M	12.5	445828.66	2229143.3	1343.34
" " M L	12.5	459063.33	2295316.7	1364.41
" " L K	12.5	488123	2440615	1409.69
" " K I	12.5	487046	2435230	1408.03
Post I j	19.76	36070.9	180354.5	843.76
" K k	19.76	42501.1	212505.5	920.69
" L l	19.76	61163.2	305816	1117.39
" M m	19.76	57260	286300	1078.88
" N n	19.76	77515.57	387577.85	1267.47
" O o	19.76	72237.4	361187	1220.81
" P p	19.76	90456.66	452283.3	1375.96
" Q q	19.76	85837	429185	1338.12
" R r	18.75	180026	900130	1794.86
Total for one half of the truss				20407.36
Amount of cast-iron				40814.72

Collecting together the parts that resist tension, and calculating their weights by the formula

$$T = \frac{Wl}{18000},$$

we obtain the following:

Tensions.

NAME.	LENGTH.	STRAIN.	FIVE TIMES THE STRAIN.	LBS. OF WROUGHT-IRON.
Bottom segment $q\,p$	12.5	51149.7	255748.5	177.60
" " $p\,o$	12.5	170484.86	852424.3	591.96
" " $o\,n$	12.5	254827.83	1274139.15	884.82
" " $n\,m$	12.5	338481.16	1692405.8	1175.28
" " $m\,l$	12.5	389968.83	1949844.15	1354.06
" " $l\,k$	12.5	441606.5	2208032.5	1533.36
" " $k\,i$	12.5	462905.16	2314525.8	1607.31
Counter-tie $q\,P$	26.5	388.9	1944.5	2.88
" " $p\,O$	26.5	2722.4	13612	20.14
" " $o\,N$	26.5	6579.6	32898	48.09
" " $n\,M$	26.5	11833.4	59167	87.56
" " $m\,L$	26.5	18449.1	92245.5	136.52
" " $l\,K$	26.5	26160.8	130804	193.58
" " $k\,I$	26.5	35035	175175	259.25
" " $j\,H$	26.5	28369.1	141845.5	209.92
Tie $j\,L$	26.5	53838.8	269194	398.39
" $k\,M$	26.5	47639.6	238198	352.52
" $l\,N$	26.5	79218.2	396091	586.19
" $m\,O$	26.5	75267.6	376338	556.96
" $n\,P$	26.5	105750.3	528751.5	782.52
" $o\,Q$	26.5	104956.2	524781	776.64
" $p\,R$	26.5	136601.4	683007	1010.81
" $q\,R$	19.76	88694.5	443472.5	486.94
Total				13233.90
Multiply by 2 for the other half of the truss				26467.80
Bottom segment $j\,i$	12.5	484894	2424470	1683.66
Amount of wrought-iron				28151.46

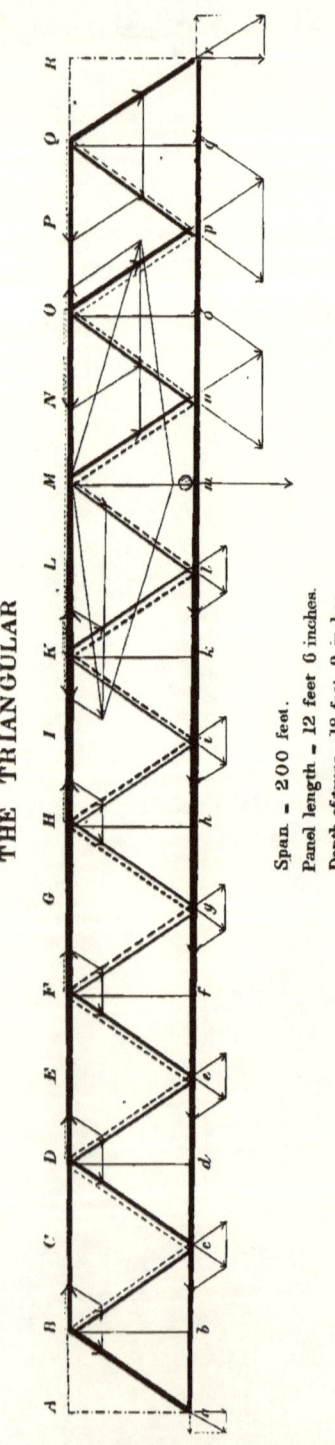

THE TRIANGULAR

Span = 200 feet.
Panel length = 12 feet 6 inches.
Depth of truss = 18 feet 9 inches.

THE TRIANGULAR TRUSS.

This truss is a modification of the truss known in England as the Warren girder, and quite extensively used there and in India. The true Warren girder is a series of wrought-iron equilateral triangles with vertices alternately up and down, the bases forming alternately the Bottom and Top Chords. In adapting it proportionally to long spans, the sides of the triangles become so large that it is necessary to introduce vertical ties to support the middle points of the segments of the Lower Chord. Sometimes posts are likewise introduced to support the middle points of the segments of the Top Chord. We have calculated the bridge without these posts, and also with them, assuming that their office is merely to diminish the amount of metal in the segments of the Top by halving their lengths. These posts will have to sustain strains of compression while preventing the joints to which they are attached from sinking, and of extension while preventing them from rising.

In the Warren girder the parts are made of wrought-iron, because the combination is such that most of the parts change their strains from compression to tension, and *vice versa* as the train crosses the bridge. This sudden reversal of strains cannot but be a very severe test for any kind of material, and to avoid it the Warren girder, as modified in this country for long spans, has its struts made hollow, with wrought-iron ties passing through their cavities. Thus each brace has a dual composition, each strain coming on the material best adapted to resist it. In the drawing the braces are drawn with both red and black lines to show this arrangement more clearly.

The intermediate supporting posts will be made like the struts, hollow, with wrought-iron vertical ties inside.

Assumed Weights and Dimensions.

$a\,r$ .... $= 200'$
No. of panels .... $= 16$
Panel length .... $= 12'\ 6''$
Depth of truss .... $= 18'\ 9''$
Panel weight of engine .... $= 17600 = w' + e$
    "   "   tender .... $= 16160 = w' + t$
    "   "   cars .... $= 13152 = w'$
    "   "   bridge .... $= 9375 = w''$

$e = 4448$
$t = 3008$

Transmission of Strains.

The same principles in regard to the transmission of balanced and unbalanced weights hold in this bridge as in the preceding.

A weight at m is at once transmitted to M, and there is resolved into components along Ma and Mr exactly as an unbalanced weight in the Jones bridge. The only difference is that the tensions at n and p are each $\frac{4}{3} \times \frac{14}{8} W$, that at r being $\frac{4}{3} \times \frac{14}{8} W$. The compressions at O and Q are each $\frac{4}{3} \times \frac{14}{8} W$, the resolution at O being between OM and Op.

An unbalanced weight at l is at once resolved along the lines lA and lR in the same way as in the Murphy-Whipple truss. The compressions generated at M, O, and Q are each $\frac{4}{3} \times \frac{10}{8} W$. The tensions at n and p are $\frac{4}{3} \times \frac{10}{8} W$, and at r $\frac{4}{3} \times \frac{10}{8} W$.

From Bridge Weight only.

Under the action of the balanced and constant bridge weights, we see that the braces are alternately ties and struts. We will see afterwards how strains of compression come upon the ties, and of extension on the struts.

Tensions on Ties.

on iK .... $\frac{1}{2} w'' \times \frac{1}{3} \sqrt{13}$
on lM .... $\frac{5}{6} w'' \times \frac{1}{3} \sqrt{13}$
on nO .... $\frac{9}{8} w'' \times \frac{1}{3} \sqrt{13}$
on pQ .... $\frac{13}{8} w'' \times \frac{1}{3} \sqrt{13}$

Compressions on Struts.

on Kl .... $\frac{3}{8} w'' \times \frac{1}{3} \sqrt{13}$
on Mn .... $\frac{7}{8} w'' \times \frac{1}{3} \sqrt{13}$
on Op .... $\frac{11}{8} w'' \times \frac{1}{3} \sqrt{13}$
on Qr .... $\frac{13}{8} w'' \times \frac{1}{3} \sqrt{13}$

Tensions on Bottom Chord.

on rp .... $\frac{13}{8} w''$
on pn .... $\frac{24}{8} w''$
on nl .... $\frac{33}{8} w''$
on li .... $\frac{40}{8} w''$

Tensions on Vertical Ties.

on kK .... w''
on mM .... w''
on oO .... w''
on qQ .... w''

Compressions on Top Chord.

on QO .... $\tfrac{5}{3}w''$
on OM .... $\tfrac{4}{3}w''$
on MK .... $\tfrac{3}{3}w''$
on KH .... $\tfrac{4}{3}w''$

Effect of Moving Load.

The effect of running a train on this bridge is the same as on any other. As the engine passes the middle point, the weights in succession balance on either side and are transmitted entire to their nearest abutments. As the inclined parts of this bridge do double duty, it is necessary to point out in advance how the braces act. It is as follows :

aB .... as Strut.
Bc .... as Counter-strut and Tie.
cD .... as Strut and Counter-tie.
De .... as Counter-strut and Tie.
eF .... as Strut and Counter-tie.
Fg .... as Counter-strut and Tie.
gH .... as Strut and Counter-tie.
Hi .... as Counter-strut and Tie.
iK .... as Counter-strut and Tie.
Kl .... as Strut and Counter-tie.
lM .... as Counter-strut and Tie.
Mn .... as Strut and Counter-tie.
nO .... as Counter-strut and Tie.
Op .... as Strut and Counter-tie.
pQ .... as Counter-strut and Tie.
Qr .... as Strut.

They undergo their maximum strains under the same circumstances as in previous bridges, namely :

Each *Counter-tie* is under its maximum direct tension when the head of the train is at its foot, the train coming from the *nearer* abutment.

Each *Tie* is under its maximum direct tension when the head of the train is at its foot, the train coming from the *further* abutment.

Each *Counter-strut* is under its maximum direct compression when the head of the train is under its top, the train coming from the *nearer* abutment.

Each *Strut* is under its maximum direct compression when the head of the train is under its top, the train coming from the *further* abutment.

Maxima Direct Tensions on Counter-Ties.

on $c\,D$ .... $(\frac{1}{18}w' + \frac{1}{18}e\qquad) \times \frac{1}{3}\sqrt{13} = 3300 \times \frac{1}{3}\sqrt{13} = 3966.1$

on $e\,F$ .... $(\frac{10}{9}w' + \frac{7}{18}e + \frac{7}{18}t) \times \frac{1}{3}\sqrt{13} = 10730 \times \frac{1}{3}\sqrt{13} = 12895.9$

on $g\,H$ .... $(\frac{24}{9}w' + \frac{11}{6}e + \frac{11}{18}t) \times \frac{1}{3}\sqrt{13} = 21636 \times \frac{1}{3}\sqrt{13} = 26003.2$

Maxima Direct Tensions on Ties.

on $i\,K$ .... $(\frac{1}{4}w'' + \frac{26}{18}w' + \frac{19}{6}e + \frac{11}{4}t) \times \frac{1}{3}\sqrt{13} = 40517.5 \times \frac{1}{3}\sqrt{13} = 48696$

on $l\,M$ .... $(\frac{9}{4}w'' + \frac{45}{18}w' + \frac{23}{6}e + \frac{13}{4}t) \times \frac{1}{3}\sqrt{13} = 76749.5 \times \frac{1}{3}\sqrt{13} = 92241.4$

on $n\,O$ .... $(\frac{9}{2}w'' + \frac{70}{18}w' + \frac{23}{6}e + \frac{15}{4}t) \times \frac{1}{3}\sqrt{13} = 116269.5 \times \frac{1}{3}\sqrt{13} = 139738.6$

on $p\,Q$ .... $(\frac{15}{2}w'' + \frac{100}{18}w' + \frac{27}{6}e + \frac{17}{4}t) \times \frac{1}{3}\sqrt{13} = 159077.5 \times \frac{1}{3}\sqrt{13} = 191187.4$

Maxima Direct Compressions on Counter-Struts.

on $B\,c$ .... $(\frac{1}{18}w' + \frac{1}{18}e\qquad) \times \frac{1}{3}\sqrt{13} = 1100 \times \frac{1}{3}\sqrt{13}$

on $D\,e$ .... $(\frac{5}{18}w' + \frac{5}{18}e + \frac{1}{18}t) \times \frac{1}{3}\sqrt{13} = 6510 \times \frac{1}{3}\sqrt{13}$

on $F\,g$ .... $(\frac{13}{9}w' + \frac{7}{6}e + \frac{5}{18}t) \times \frac{1}{3}\sqrt{13} = 15772 \times \frac{1}{3}\sqrt{13}$

on $H\,i$ .... $(\frac{24}{9}w' + \frac{14}{9}e + \frac{7}{18}t) \times \frac{1}{3}\sqrt{13} = 28322 \times \frac{1}{3}\sqrt{13}$

Maxima Direct Compressions on Struts.

on $K\,l$ .... $(\frac{3}{2}w'' + \frac{45}{18}w' + \frac{17}{6}e + \frac{13}{4}t) \times \frac{1}{3}\sqrt{13} = 58222.5 \times \frac{1}{3}\sqrt{13}$

on $M\,n$ .... $(\frac{7}{2}w'' + \frac{66}{18}w' + \frac{21}{6}e + \frac{15}{4}t) \times \frac{1}{3}\sqrt{13} = 96098.5 \times \frac{1}{3}\sqrt{13}$

on $O\,p$ .... $(\frac{11}{2}w'' + \frac{94}{18}w' + \frac{25}{6}e + \frac{17}{4}t) \times \frac{1}{3}\sqrt{13} = 137262.5 \times \frac{1}{3}\sqrt{13}$

on $Q\,r$ .... $(\frac{15}{2}w'' + \frac{119}{18}w' + \frac{29}{6}e + \frac{19}{4}t) \times \frac{1}{3}\sqrt{13} = 181714.5 \times \frac{1}{3}\sqrt{13}$

Maxima Tensions on Bottom Chord.

on $p\,r$ .... $\frac{15}{4}w'' + \frac{30}{9}w' + \frac{35}{6}e + \frac{50}{18}t = 121143$

on $n\,p$ .... $\frac{31}{4}w'' + \frac{78}{9}w' + \frac{19}{6}e + \frac{150}{18}t = 309479$

on $l\,n$ .... $\frac{54}{4}w'' + \frac{110}{9}w' + \frac{23}{6}e + \frac{175}{18}t = 428695.66$

on $i\,l$ .... $\frac{63}{4}w'' + \frac{129}{9}w' + \frac{24}{6}e + \frac{178}{18}t = 485967$

Tensions on Vertical Ties.

on $k\,K$ .... $w'' + w' = 22527$

on $m\,M$ .... $w'' + w' = 22527$

on $o\,O$ .... $w'' + w' = 22527$

on $q\,Q$ .... $w'' + w' = 22527$

Maxima Compressions on Top Chord.

As in other trusses, these will be found when the entire bridge is covered—they are as follows:

on HK .... $\frac{5.4}{3} w'' + \frac{5.4}{3} w' + \frac{1.2}{3} e + \frac{2.3}{3} t = 492042.66$
on KM .... $\frac{5.0}{3} w'' + \frac{5.0}{3} w' + \frac{1.3}{3} e + \frac{3.3}{3} t = 464873.63$
on MO .... $\frac{3.3}{3} w'' + \frac{3.3}{3} w' + \frac{1.3}{3} e + \frac{4.3}{3} t = 377632$
on OQ .... $\frac{3.3}{3} w'' + \frac{3.3}{3} w' + \frac{2.1}{3} e + \frac{2.3}{3} t = 224302.66$

Extra Strains.

Any tendency of the joints of the Top Chord to fall below the line of the chord must bring extra strains on the struts. Any tendency to rise above the line of the Top, that is not counteracted by an equal or greater transmitted weight acting downward, must be met by the vertical ties or by the ties proper. The former are fastened at their lower extremities to the Bottom Chord, which resists flexure ; and being, in direction, directly opposed to the upward force, we may conclude that they, rather than the two inclined ties, will resist it.

Total Tensions on Vertical Ties.

The only extra strain comes on kK ; because, when the entire bridge is loaded, the direct strain on kK is 22527, the compression at K is 464873.33, $\frac{1}{17}$th of which is 27345.5, and the transmitted downward weight at K (other than the w'' that comes through Kk) is $\frac{1}{2} w'' = 4687.5$. Subtracting this from the upward force, we get 22658, which must be sustained by kK.

At M the transmitted weight is $\frac{4}{3} w''$, much more than the $\frac{1}{17}$th of the Top compression, and hence there is no extra strain on mM. For stronger reasons there is none on oO, and, as there is no rising or falling at Q, of course there is none on qQ.

Total Compressions on Counter-Struts and Struts.

The greatest total compressions on the struts will not necessarily be when they are under their maxima direct compressions. The amount to be added to each strut is one-thirty-fourth of the then compression of the Top at the joint in question, multiplied by the secant of the inclination. We will compute the sum of the direct strain and the extra strain when the engine is at different positions, and take the sum that is greatest.

No extra strain can come on Ba and Bc, nor on Qr and Qp, and we will therefore neglect them.

Total Strains on Braces.

Name.	Position of Engine.	Direct Strain.	Compression at Joint.		Total Strain.
De	at d	$6510 \times \frac{1}{3} \sqrt{13}$	D	136526.66	$10525.5 \times \frac{1}{3} \sqrt{13}$
"	at e	$6330 \times \frac{1}{3} \sqrt{13}$	D	152446.66	$10813.7 \times \frac{1}{3} \sqrt{13}$
"	at f	$5872 \times \frac{1}{3} \sqrt{13}$	D	165265.33	$10732.7 \times \frac{1}{3} \sqrt{13}$
"	at h	$4932 \times \frac{1}{3} \sqrt{13}$	D	210270.66	$11116.4 \times \frac{1}{3} \sqrt{13}$
Fg	at f	$15772 \times \frac{1}{3} \sqrt{13}$	F	264443	$23549.7 \times \frac{1}{3} \sqrt{13}$
"	at g	$15036 \times \frac{1}{3} \sqrt{13}$	F	287888	$23503.3 \times \frac{1}{3} \sqrt{13}$
"	at l	$12330 \times \frac{1}{3} \sqrt{13}$	F	345712	$22498 \times \frac{1}{3} \sqrt{13}$
Hi	at h	$28322 \times \frac{1}{3} \sqrt{13}$	H	351051	$38647 \times \frac{1}{3} \sqrt{13}$
"	at i	$27030 \times \frac{1}{3} \sqrt{13}$	H	391094	$38533 \times \frac{1}{3} \sqrt{13}$
Kl	at k	$58222.5 \times \frac{1}{3} \sqrt{13}$	K	364140	$68932.5 \times \frac{1}{3} \sqrt{13}$
Mn	at m	$96098.5 \times \frac{1}{3} \sqrt{13}$	M	318763	$105473.9 \times \frac{1}{3} \sqrt{13}$
Op	at o	$137262.5 \times \frac{1}{3} \sqrt{13}$	O	201767	$143196.8 \times \frac{1}{3} \sqrt{13}$

Selecting the greatest strains, and including the other braces, we have

on Bc .... $1100 \times \frac{1}{3} \sqrt{13} = 1322$
on De .... $11116.4 \times \frac{1}{3} \sqrt{13} = 13360.25$
on Fg .... $23549.7 \times \frac{1}{3} \sqrt{13} = 28303.2$
on Hi .... $38647 \times \frac{1}{3} \sqrt{13} = 46448.$
on Kl .... $68932.5 \times \frac{1}{3} \sqrt{13} = 82846.5$
on Mn .... $105473.9 \times \frac{1}{3} \sqrt{13} = 126763.8$
on Op .... $143196.8 \times \frac{1}{3} \sqrt{13} = 172101$
on Qr .... $181714.5 \times \frac{1}{3} \sqrt{13} = 218393.9$

Collecting together the parts that resist compression, and calculating their weights by the formula

$$P = 0.00459061 \times W^{\frac{1}{1.78}} \times l^{\frac{1.78}{0.78}},$$

we obtain the following:

Compressions (1st Case).

NAME.	LENGTH.	STRAIN.	FIVE TIMES THE STRAIN.	LBS. OF CAST-IRON.
Top segment QO	25	224302.66	1121513.3	3489.28
" " OM	25	377632	1888160	4603.25
" " MK	25	464873.33	2324366.7	5177.11
Counter-strut Qp	22.53	1322	6610	186.60
" " On	22.53	13360.3	66801.5	638.71
" " Ml	22.53	28303.2	141516	952.11
" " Ki	22.53	46448	232240	1239.13
Strut Kl	22.53	82846.5	414232.5	1085.75
" Mn	22.53	126763.8	633819	2113.72
" Op	22.53	172101	860505	2487.02
" Qr	22.53	218393.9	1091969.5	2823.00
Total				25395.78
Multiply by 2 for the other half of the truss				50791.56
Top segment KH	25	492042.66	2460213.3	5299.19
Amount of cast-iron				56090.75

Collecting together the parts that resist tension, and calculating their weights by the formula

$$T = \frac{Wl}{18000},$$

we obtain the following:

Tensions (1st Case).

NAME.	LENGTH.	STRAIN.	FIVE TIMES THE STRAIN.	LBS. OF WROUGHT-IRON.
Bottom segment rp	25	121143	605715	841.27
" " pn	25	309479	1547395	2149.16
" " nl	25	428695.66	2143478.3	2977.05
" " li	25	485967	2429835	3374.77
Counter-tie Op	22.53	3966.1	19830.5	24.83
" Mn	22.53	12895.9	64474.5	80.72
" Kl	22.53	26003.2	130016	162.77
Tie Ki	22.53	48696	243480	304.82
" Ml	22.53	92241.4	461207	557.40
" On	22.53	139738.6	698693	874.71
" Qp	22.53	191187.4	955937	1196.76
Vertical tie Kk	18.75	22658	113290.5	118.01
" Ll	.....	....	....	....
" Mm	18.75	22527	112635	117.33
" Nn	.....	....	....	....
" Oo	18.75	22527	112635	117.33
" Pp	.....	....	....	....
" Qq	18.75	22527	112635	117.33
Total for one-half of the truss....				13014.26
Amount of wrought-iron.......				26028.52

Introducing Intermediate Posts.

Our previous calculations have been based upon the supposition that each segment of the Top Chord was of the length of two panels. If, now, we cut these long top segments into two shorter ones, each of the length of one panel, we must introduce a member to keep the joint in question from rising above, or falling below, the line of the Top.

A simple post will prevent the joint from falling, and a vertical tie will prevent it from rising. Following the same plan with these as with the braces, we will make the post hollow and will insert the tie in the cavity.

The strain on either post or tie will be, as in similar cases, one-seventeenth of the top compression at the joint.

We found before, in discussing the Fink truss, that there is 87 per cent. more metal in a long post than there is in two short ones, under the same strain and with but half the length. This saving in the Top Chord is in some measure made up by the weight of the new member.

Collecting together our new compressions, and calculating them as before, we get: .

COMPRESSIONS (2D CASE).

NAME.	LENGTH.	STRAIN.	FIVE TIMES THE STRAIN.	LBS. OF CAST-IRON.
Top segment QP	12.5	224302.66	1121513.30	932.18
" " PO	12.5	224302.66	1121513.30	932.18
" " ON	12.5	377632	1888160	1229.81
" " NM	12.5	377632	1888160	1229.81
" " ML	12.5	464873.33	2324366.65	1383.09
" " LK	12.5	464873.33	2324366.65	1383.09
" " KI	12.5	492042.66	2460213.3	1415.70
Post Ll	18.75	27345.5	136727.5	658.69
" Nn	18.75	22213.6	111068	589.75
" Pp	18.75	13194.3	65971.5	447.02
Counter-strut Qp	22.53	1322	6610	186.60
" " On	22.53	13360.3	66801.5	688.71
" " Ml	22.53	28303.2	141516	952.11
" " Ki	22.53	46448	232240	1239.13
Strut Kl	22.53	82846.5	414232.5	1685.75
" Mn	22.53	126763.8	633819	2113.72
" Op	22.53	172101	860505	2487.02
" Qr	22.53	218393.9	1091969.5	2823.00
Total				22327.35
Multiply by 2 for the other half of the truss				44654.70
Post Ii	18.75	28043.7	144718.5	678.89
Amount of cast-iron				45333.50

Collecting together the new tensions, and calculating the weights as before, we get:

Tensions (2d Case).

NAME.	LENGTH.	STRAIN.	FIVE TIMES THE STRAIN.	LBS. OF WROUGHT-IRON.
Bottom segment $r\,p$	25.	121143	605715	841.27
" " $p\,n$	25.	300479	1547395	2149.16
" " $n\,l$	25.	428695.66	2143478.3	2977.05
" " $l\,i$	25.	485067	2429835	3374.77
Counter-tie $O\,p$	22.53	3966.1	19830.5	24.83
" " $M\,n$	22.53	12895.9	64474.5	80.72
" " $K\,l$	22.53	26003.2	130016	162.77
Tie $K\,i$	22.53	48696	243480	304.82
" $M\,l$	22.53	92241.4	461207	557.40
" $O\,n$	22.53	139738.6	698693	874.71
" $Q\,p$	22.53	191187.4	955937	1196.76
Vertical tie $K\,k$	18.75	22658	113290	118.01
" " $L\,l$	18.75	27345.5	136727.5	142.42
" " $M\,m$	18.75	22527	112635	117.33
" " $N\,n$	18.75	22213.6	111068	115.70
" " $O\,o$	18.75	22527	112635	117.33
" " $P\,p$	18.75	13194.3	65971.5	68.72
" " $Q\,q$	18.75	22527	112635	117.33
Total				13341.10
Multiply by 2 for the other half of the truss				26682.20
Vertical tie $I\,i$	18.75	28943.7	144718.5	150.75
Amount of wrought-iron				26832.95

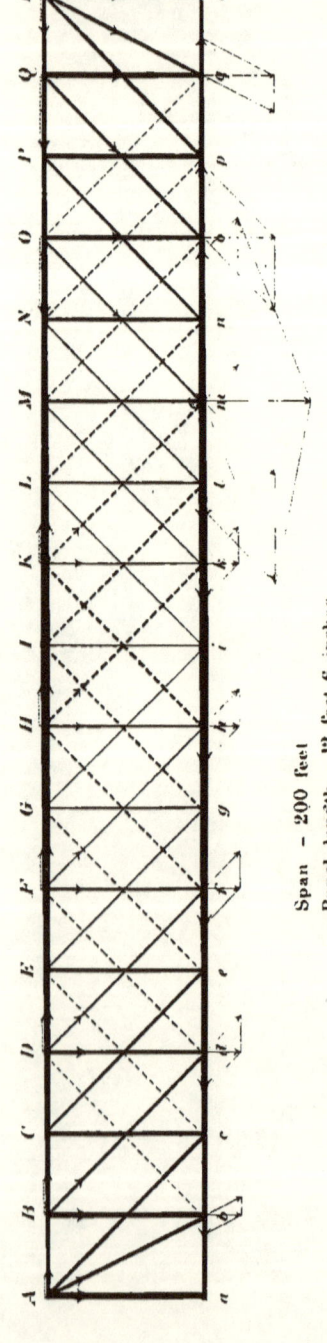

LINVILLE

Span = 200 feet
Panel length = 12 feet 6 inches
Depth of truss = 25 feet

THE LINVILLE TRUSS.

This truss is a modification of one of the Whipple bridges, the Murphy-Whipple being a modification of another. It differs from the Murphy-Whipple bridge in having what is called a double intersection, the run of the braces being two panel lengths instead of one; each brace, therefore, crosses its adjacent post. The diagram shows its construction clearly. The posts are vertical and for sake of comparison we will assume them to be made of cast-iron; the braces are inclined and are made of wrought-iron.

No calculations have been made upon the original Whipple bridges, for, though they were great improvements in scientific bridge-building at the time they were invented, they would require before discussion the introduction of various parts that would make them either the Murphy-Whipple or the Linville modifications.

Assumed Weight and Dimensions.

$a\,r$ .... $= 200'$
Number of panels.................................. $= 16$
Panel length.. $= 12'\ 6''$
Depth of truss...................................... $= 25'$
Panel weight of engine .... $= 17000 = w' + e$
   "   "   " tender .... $= 16160 = w' + t$
   "   "   " cars .... $= 13152 = w'$
   "   "   " bridge .... $=\ \ 9375 = w''$
$$e = 4448$$
$$t = 3008$$

We have made the depth of the truss twenty-five feet, as in no other way could we secure panel lengths of twelve and a half feet, whilst conforming to the practice of the builders of the truss. The increase in depth of truss is no loss, as what is lost by increasing the lengths of posts and braces is made up by the diminution of the strains. The comparison with other trusses will be more just if we retain the same panel length and number of panels, than if we change them in order to secure the same depth of truss.

Transmission of Strains.

Balanced weights are transmitted, as in all the preceding trusses, wholly to the nearest abutment.

An unbalanced weight at m, for instance, is at once resolved into two components along the theoretical lines $m\,A$ and $m\,R$. Following the component along $m\,R$, we find it resolved into one along $m\,r$, which diminishes the bottom tension produced by the component along mA, and one producing tension on mO. This last produces compression on the Top Chord at O, compression on the post Oo, tension on the Bottom Chord at o, tension on oQ, compression on the Top Chord at Q, compression on Qq, tension on the Bottom Chord at q, tension on qR, compression on the Top Chord at R, and compression on the end post Rr. The other component, along mA, would produce similar effects, which are shown on the diagram of the truss.

From Bridge Weight only.

Tensions on Ties.

$$
\begin{aligned}
\text{on } iL &\;\ldots\; \tfrac{1}{2} w'' \times \sqrt{2} \\
\text{on } kM &\;\ldots\; w'' \times \sqrt{2} \\
\text{on } lN &\;\ldots\; \tfrac{3}{2} w'' \times \sqrt{2} \\
\text{on } mO &\;\ldots\; 2 w'' \times \sqrt{2} \\
\text{on } nP &\;\ldots\; \tfrac{5}{2} w'' \times \sqrt{2} \\
\text{on } oQ &\;\ldots\; 3 w'' \times \sqrt{2} \\
\text{on } pR &\;\ldots\; \tfrac{7}{2} w'' \times \sqrt{2} \\
\text{on } qR &\;\ldots\; 4 w'' \times \tfrac{1}{2}\sqrt{5}
\end{aligned}
$$

Tensions on Bottom Chord.

$$
\begin{aligned}
\text{on } rq &\;\ldots\; 0 \\
\text{on } qp &\;\ldots\; 2 w'' \\
\text{on } po &\;\ldots\; \tfrac{11}{2} w'' \\
\text{on } on &\;\ldots\; \tfrac{15}{2} w'' \\
\text{on } nm &\;\ldots\; \tfrac{21}{2} w'' \\
\text{on } ml &\;\ldots\; \tfrac{23}{2} w'' \\
\text{on } lk &\;\ldots\; \tfrac{27}{2} w'' \\
\text{on } ki &\;\ldots\; \tfrac{27}{4} w''
\end{aligned}
$$

Compressions on Posts.

$$
\begin{aligned}
\text{on } Ii &\;\ldots\; 0 \\
\text{on } Kk &\;\ldots\; 0 \\
\text{on } Ll &\;\ldots\; \tfrac{1}{2} w'' \\
\text{on } Mm &\;\ldots\; w'' \\
\text{on } Nn &\;\ldots\; \tfrac{3}{2} w'' \\
\text{on } Oo &\;\ldots\; 2 w'' \\
\text{on } Pp &\;\ldots\; \tfrac{5}{2} w'' \\
\text{on } Qq &\;\ldots\; 3 w'' \\
\text{on } Rr &\;\ldots\; \tfrac{7}{2} w''
\end{aligned}
$$

IRON TRUSS BRIDGES FOR RAILROADS.   105

Compressions on Top Chord.

on $R\ Q$ .... $1\frac{1}{2}\ w''$
on $Q\ P$ .... $1\frac{7}{9}\ w''$
on $P\ O$ .... $2\frac{2}{9}\ w''$
on $O\ N$ .... $2\frac{5}{9}\ w''$
on $N\ M$ .... $2\frac{7}{9}\ w''$
on $M\ L$ .... $3\frac{1}{9}\ w''$
on $L\ K$ .... $3\frac{2}{9}\ w''$
on $K\ I$ .... $3\frac{3}{9}\ w''$

EFFECT OF MOVING LOAD.

In this bridge the ties and counter-ties sustain the maxima direct tensions when the head of the train is at the same places as in the Murphy-Whipple. The top and bottom chords and end posts are evidently most strained when the entire bridge is covered by the train. The intermediate posts sustain the greatest compressive strains when the ties attached to their upper extremities are under their greatest tensile strains.

Maxima Direct Tensions on Counter-Ties.

on $a\ B$ 0
on $a\ C$ 0
on $b\ D$ .... $(\frac{1}{10} w' + \frac{1}{10} e)$ $\times \sqrt{2} =$ $1100 \times \sqrt{2} =$ 1555.6
on $c\ E$ ... $(\frac{1}{10} w' + \frac{2}{10} e)$ $\times \sqrt{2} =$ $2200 \times \sqrt{2} =$ 3111.3
on $d\ F$ .... $(\frac{1}{10} w' + \frac{3}{10} e + \frac{1}{10} t) \times \sqrt{2} =$ $4310 \times \sqrt{2} =$ 6095.3
on $e\ G$ ... $(\frac{1}{10} w' + \frac{4}{10} e + \frac{2}{10} t) \times \sqrt{2} =$ $6420 \times \sqrt{2} =$ 9079.3
on $f\ H$ .... $(\frac{1}{10} w' + \frac{5}{10} e + \frac{3}{10} t) \times \sqrt{2} =$ $9352 \times \sqrt{2} =$ 13225.7
on $g\ I$ .... $(1\frac{2}{5} w' + \frac{6}{10} e + \frac{4}{10} t) \times \sqrt{2} =$ $12284 \times \sqrt{2} =$ 17372.2
on $h\ K$ .... $(1\frac{4}{5} w' + \frac{7}{10} e + \frac{5}{10} t) \times \sqrt{2} =$ $16038 \times \sqrt{2} =$ 22681.2

Maxima Direct Tensions on Ties.

on $i\ L$ .... $(\frac{1}{2} w'' + \frac{22}{9} w' + \frac{8}{10} e + \frac{6}{10} t) \times \sqrt{2} =$ $24479.5 \times \sqrt{2} =$ 34619.2
on $k\ M$ .... $(\quad w'' + \frac{22}{9} w' + \frac{9}{10} e + \frac{7}{10} t) \times \sqrt{2} =$ $33743 \times \sqrt{2} =$ 47719.8
on $l\ N$ .... $(\frac{3}{2} w'' + \frac{22}{9} w' + \frac{10}{10} e + \frac{8}{10} t) \times \sqrt{2} =$ $43006.5 \times \sqrt{2} =$ 60820.4
on $m\ O$ .... $(2 w'' + \frac{22}{9} w' + \frac{11}{10} e + \frac{9}{10} t) \times \sqrt{2} =$ $53092 \times \sqrt{2} =$ 75083.4
on $n\ P$ .... $(\frac{5}{2} w'' + \frac{22}{9} w' + \frac{12}{10} e + \frac{10}{10} t) \times \sqrt{2} =$ $63177.5 \times \sqrt{2} =$ 89346.5
on $o\ Q$ .... $(3 w'' + \frac{22}{9} w' + \frac{13}{10} e + \frac{11}{10} t) \times \sqrt{2} =$ $74085 \times \sqrt{2} =$ 104772
on $p\ R$ .... $(\frac{7}{2} w'' + \frac{22}{9} w' + \frac{14}{10} e + \frac{12}{10} t) \times \sqrt{2} =$ $84992.5 \times \sqrt{2} =$ 120197.6
on $q\ R$ .... $(4 w'' + \frac{22}{9} w' + \frac{15}{10} e + \frac{13}{10} t) \times \frac{1}{2} \sqrt{5} =$ $96722 \times \frac{1}{2} \sqrt{5} =$ 108138.5

Maxima Tensions on Bottom Chord.

With the head of the train at q we get

on qr 0
on pq .... $2\,w'' + 2\,w' + \frac{13}{3}e + \frac{13}{3}t = 48083$
on op .... $\frac{11}{3}w'' + \frac{11}{3}w' + \frac{34}{3}e + \frac{34}{3}t = 132519.5$
on no .... $\frac{17}{3}w'' + \frac{17}{3}w' + \frac{53}{3}e + \frac{53}{3}t = 201702.5$
on mn .... $\frac{22}{3}w'' + \frac{22}{3}w' + \frac{31}{3}e + \frac{73}{3}t = 258968$
on lm .... $\frac{26}{3}w'' + \frac{26}{3}w' + \frac{32}{3}e + \frac{61}{3}t = 303180$
on kl .... $\frac{29}{3}w'' + \frac{29}{3}w' + \frac{70}{6}e + \frac{40}{3}t = 335662.5$
on ik .... $\frac{31}{3}w'' + \frac{31}{3}w' + \frac{73}{3}e + \frac{94}{3}t = 357347.5$

Maxima Direct Compressions on Posts.

on Ii .... $\frac{13}{8}w' + \frac{4}{15}e + \frac{4}{15}t = 12284$
on Kk .... $\frac{16}{8}w' + \frac{4}{15}e + \frac{4}{15}t = 16038$
on Ll .... $\frac{1}{2}w'' + \frac{26}{8}w' + \frac{4}{15}e + \frac{4}{15}t = 24479.5$
on Mm .... $w'' + \frac{25}{8}w' + \frac{7}{15}e + \frac{7}{15}t = 33743$
on Nn .... $\frac{3}{2}w'' + \frac{32}{8}w' + \frac{10}{15}e + \frac{4}{15}t = 43006.5$
on Oo .... $2\,w'' + \frac{40}{8}w' + \frac{14}{15}e + \frac{7}{15}t = 53092$
on Pp .... $\frac{5}{2}w'' + \frac{12}{8}w' + \frac{13}{15}e + \frac{13}{15}t = 63177.5$
on Qq .... $3\,w'' + \frac{56}{8}w' + \frac{17}{15}e + \frac{11}{15}t = 74085$
on Rr .... $\frac{16}{2}w'' + \frac{14}{2}w' + \frac{27}{8}e + \frac{24}{8}t = 181714.5$

Maxima Compressions on Top Chord.

As before, the entire bridge is loaded. They are—

on IK .... $\frac{22}{3}w'' + \frac{22}{3}w' + \frac{32}{3}e + \frac{91}{3}t = 370761$
on KL .... $\frac{22}{3}w'' + \frac{22}{3}w' + \frac{31}{3}e + \frac{73}{3}t = 371603$
on LM .... $\frac{21}{3}w'' + \frac{21}{3}w' + \frac{34}{3}e + \frac{61}{3}t = 361647.5$
on MN .... $\frac{22}{3}w'' + \frac{22}{3}w' + \frac{31}{3}e + \frac{31}{3}t = 339962.5$
on NO .... $\frac{26}{3}w'' + \frac{26}{3}w' + \frac{31}{3}e + \frac{37}{3}t = 306728$
on OP .... $\frac{22}{3}w'' + \frac{22}{3}w' + \frac{43}{3}e + \frac{34}{3}t = 261952$
on PQ .... $\frac{17}{3}w'' + \frac{17}{3}w' + \frac{33}{6}e + \frac{64}{3}t = 203378.5$
on QR .... $\frac{11}{3}w'' + \frac{11}{3}w' + \frac{32}{6}e + \frac{31}{3}t = 133353.5$

EXTRA STRAINS.

The tendency of the joints of the Top Chord to get out of line by upward divergence must be met by the braces at the joints. The tendency to divergence downwards must be met and resisted by the posts.

Total Tensions on Counter-Ties and Ties.

The upward tendency of the joints of the Top Chord is, in a great measure, counteracted by the transmitted weights. The maxima values of these upward forces on a single brace are $\frac{1}{34}$th the maxima compressions of the Top Chord. These are

at O or D .... $\frac{1}{34} \times 261952 = 7704.5$
at N or E .... $\frac{1}{34} \times 306728 = 9021.4$
at M or F .... $\frac{1}{34} \times 339962 = 9998.9$
at L or G .... $\frac{1}{34} \times 361647 = 10636.7$
at K or H .... $\frac{1}{34} \times 370761 = 10904.7$
at I .... $\frac{1}{34} \times 370770 = 10905$ (engine at p).

At P and Q the transmitted weights are always much greater than the $\frac{1}{34}$th of the top compression, and therefore there is no tendency to rise at these points. The transmitted weights are also greater than the upward force at O, N, M and L, and therefore no extra strains can be brought on the braces that meet at these points. At I and K there are no transmitted weights (except parts of e and t) when the entire bridge is covered, and therefore the braces at these points must be strong enough to resist these upward forces; but the *ties* that meet at these points are already strong enough to resist these upward forces. We need therefore only strengthen the *counter-tie* $m K$, or its equal $f H$, and $l I$, or $g I$. Hence,

on $f H$ .... $10904.7 \times \sqrt{2} = 15421.6$
on $g I$ .... $10905 \times \sqrt{2} = 15422$

Total Compressions on Posts.

The posts must be able to sustain $\frac{1}{17}$th the greatest local top compression when they are under their maxima direct compressions. They remain under their maxima compressions after the engine has passed to the panel beyond them; and sometimes there is but little diminution in their direct compressions after the engine has gone forward more than one panel. Calculating the required top compressions, we have on the posts:

Engine at k — on $I i$ .... $10992 + \frac{1}{17} \times 321854 = 29924.6$
" at i — on $K k$ .... $16038 + \frac{1}{17} \times 285301 = 32820.4$
" at k — on $L l$ .... $24479.5 + \frac{1}{17} \times 289976.5 = 41537$
" at l — on $M m$ .... $33743 + \frac{1}{17} \times 283689.5 = 50430.6$
" at m — on $N n$ .... $43006.5 + \frac{1}{17} \times 265618 = 58631$
" at n — on $O o$ .... $53092 + \frac{1}{17} \times 234118 = 66863.6$
" at o — on $P p$ .... $63177.5 + \frac{1}{17} \times 188367.5 = 74257.9$
" at p — on $Q q$ .... $74085 + \frac{1}{17} \times 126722.5 = 81539.3$
on $R r$ .... $= 181714.5$

We have found no strains on rq or ab; but there is the same necessity for them that exists in the Fink, the Bollman, the Murphy-Whipple, and the Post—to prevent the pull of the drivers of the locomotive from moving the Bottom Chord.

Collecting together the parts that resist compression, and calculating their weights by the formula

$$P = 0.00459961 \times W^{\frac{1}{LTB}} \times l^{\frac{LTB}{ETB}},$$

we obtain the following:

COMPRESSIONS.

NAME.	LENGTH.	STRAIN.	FIVE TIMES THE STRAIN.	LBS. OF CAST-IRON.
Top segment RQ	12.5	133353.5	666767.5	706.93
" " QP	12.5	203378.5	1016892.5	884.86
" " PO	12.5	261952	1309760	1012.38
" " ON	12.5	306728	1533640	1101.02
" " NM	12.5	339962.5	1699812.5	1162.95
" " ML	12.5	361647.5	1808237.5	1201.84
" " LK	12.5	371603	1858015	1219.32
" " KI	12.5	370761	1853805	1217.85
Post Kk	25	32820.4	164102	1255.31
" Ll	25	41537	207685	1422.86
" Mm	25	50430.6	252153	1577.54
" Nn	25	58631	293155	1709.17
" Oo	25	66863.6	334318	1832.90
" Pp	25	74257.9	371289.5	1938.07
" Qq	25	81539.3	407696.5	2036.94
" Rr	25	181714.5	908572.5	3119.58
Total				23400.52
Multiply by 2 for the other half of the truss				46801.04
Post Ii	25	29924.6	149623	1195.13
Amount of cast-iron				47996.17

IRON TRUSS BRIDGES FOR RAILROADS. 109

Collecting together the parts that resist tension, and calculating the weights by the formula

$$T = \frac{Wl}{18000},$$

we obtain the following:

TENSIONS.

NAME.	LENGTH.	STRAIN.	FIVE TIMES THE STRAIN.	LBS. OF WROUGHT-IRON.
Bottom segment $q\,p$	12.5	48083	240415	166.95
" " $p\,o$	12.5	132519.5	662597.5	460.14
" " $o\,n$	12.5	201702.5	1008512.5	700.35
" " $n\,m$	12.5	258968	1294840	899.19
" " $m\,l$	12.5	303180	1515900	1052.71
" " $l\,k$	12.5	335662.5	1678312.5	1165.50
" " $k\,i$	12.5	357347.5	1786732.5	1240.79
Counter-tie $O\,q$	35.355	1555.6	7778	15.28
" " $N\,p$	35.355	3111.3	15556.5	30.56
" " $M\,o$	35.355	6095.3	30476.5	59.86
" " $L\,n$	35.355	9079.3	45396.5	89.17
" " $K\,m$	35.355	15421.6	77108	151.46
" " $I\,l$	35.355	15422	77110	151.46
" " $H\,k$	35.355	22681.2	113406	222.75
Tie $L\,i$	35.355	34619.2	173096	339.99
" $M\,k$	35.355	47719.8	238599	468.65
" $N\,l$	35.355	60820.4	304102	597.31
" $O\,m$	35.355	75083.4	375417	737.39
" $P\,n$	35.355	89346.5	446732.5	877.47
" $Q\,o$	35.355	104772	523860	1028.96
" $R\,p$	35.355	120197.6	600988	1180.45
" $R\,q$	27.951	108138.5	540692.5	839.60
Total for one half of the truss....				12475.79
Amount of wrought-iron................				24951.58

COMPARISON OF WEIGHTS OF BRIDGES.

Having finished the calculations for the weight of each member of each of the seven trusses examined, we will now proceed to compare the weights of bridges erected on these different plans. Having followed the same methods of investigation in all of the bridges, and having demanded the same strength from each, we may conclude that our comparisons will give the relative values of the combinations in the different trusses as far as concerns the *economy of form*. The lightest bridge will evidently be the cheapest, and probably the best.

Allowance for Mechanical Connections.

The weights that we have determined are all calculated for shorter lengths than can be used in practice. We have made no allowance for the extra material needed for joints, eyes, nuts, bolts, etc., to secure the mechanical connections of the different parts. No definite percentage for this can be given, as it depends entirely upon the practical skill of the inventor or builder of the bridge. It may be assumed, however, as varying from 10 to 20 per cent., but the discussion of its exact amount is unnecessary for the purposes of this treatise, as we assume equal skill in this matter in the builders of each truss. We will add 15 per cent. to the weights previously determined, to cover this necessity.

Allowance for Top-bracing and Flooring.

Besides the weight of the two trusses that form the ordinary railroad bridge we must include the top lateral bracing, that connects the trusses, and the roadway-bearers with the track and its appurtenances. We will then add

For iron floor beams..................................	1260	lbs. per panel of bridge.
For top lateral struts.................................	270	" " " "
" " " ties.............................	300	" " " "
For track (including rails, stringers, ties, chairs, foot-walk, etc.)..	3070	" " " "
Total...................................	4900	" " " "

which is 2450 lbs. per panel for one truss. For our bridges, that have 16 panels, we must therefore add

$$4900 \times 16 = 78400 \text{ lbs.}$$

Tabulating our results, we get:

IRON TRUSS BRIDGES FOR RAILROADS.

	Cast-Iron.	Wrought-Iron.	Total.	15 %	One Truss.	Two Trusses.
Fink............	45369.00	45890.32	91259.32	6083.95	97343.27	194686.54
Bollman.........	41920.82	93765.32	135686.14	9045.74	144731.88	289463.76
Jones	61585.72	29092.04	90677.76	6045.18	96722.94	193445.88
Murphy-Whipple...	42918.16	31093.92	74012.08	4934.14	78946.22	157892.44
Post.............	40814.72	28151.46	68966.18	4597.75	73563.93	147127.86
Triangular—1.....	56090.75	26028.52	82119.27	5474.62	87593.89	175187.78
" —2.....	45333.59	26832.95	72166.54	4811.10	76977.64	153955.28
Linville..........	47996.17	24951.58	72947.75	4863.18	77810.93	155621.86

	Bracing and Flooring.	Weight of Bridge.	Less than assumed Bridge Weight.	Difference per Panel.	Order of Lightness.		
					Cast-iron.	Wrought-Iron.	Bridge Weight.
Fink	78400	273086.54	26913.46	1682.10	5	7	7
Bollman	78400	367863.76	−67863.76	−4241.49	2	8	8
Jones	78400	271845.88	28154.12	1759.63	8	5	6
Murphy-Whipple	78400	236292.44	63707.56	3981.72	3	6	4
Post	78400	225527.86	74472.14	4654.51	1	4	1
Triangular—1	78400	253587.78	46412.22	2900.77	7	2	5
" —2	78400	232355.28	67644.72	4227.80	4	3	2
Linville	78400	234021.86	65978.14	4123.63	6	1	3

We see from this comparison that the bridges are in order of lightness as follows:

1. Post.
2. Triangular (with intermediate posts).
3. Linville.
4. Murphy-Whipple.
5. Triangular (without intermediate posts.)
6. Jones.
7. Fink.
8. Bollman.

All but the last are lighter than the originally assumed weight of bridge, and, if re-calculated with the new values of w'', would be still lighter, except Bollman's, which would require a still greater amount of metal. Or, each of the first seven bridges may be supposed to be loaded with the following additional weights per panel:

Post	4654.51
Triangular (with posts)	4227.80
Linville	4123.63
Murphy-Whipple	3981.72
Triangular (without posts)	2900.77
Jones	1759.63
Fink	1682.10

Comparing these bridges with Fink's, we find that, as they stand, each will carry the following number of pounds more per panel:

Post	2972.41
Triangular (with posts)	2545.70
Linville	2441.53
Murphy-Whipple	2299.62
Triangular (without posts)	1218.67
Jones	77.53

We will now endeavor to ascertain why these differences exist, and, with that end, we will try to find out what should be the angles of ties and struts to transmit given strains with the minimum amount of material.

BEST ANGLE FOR A PAIR OF TIES.

TO DETERMINE THE PROPER ANGLE OF INCLINATION FOR A PAIR OF WROUGHT-IRON TIES THAT THEY MAY TRANSMIT A GIVEN WEIGHT TO THE POINTS OF SUPPORT WITH THE MINIMUM AMOUNT OF METAL.

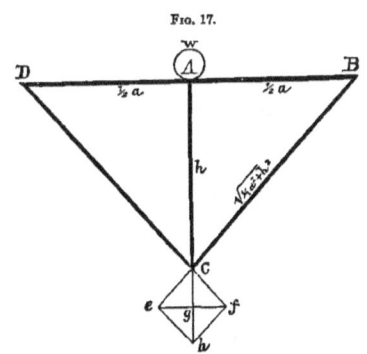

Fig. 17.

Let DB be a beam, of the length a, strengthened by the post AC and the ties CD and CB, and with the weight W at A. The weight is at once transmitted through AC to C, whence it is carried to D and B by the ties CD and CB. We wish to ascertain what should be the inclination of CB to the vertical that it may transmit its half of W to B with the minimum amount of material.

W at C is represented by Ch, which is resolved into Ce and Cf. From ABC and Ceg we get

$$Cg : AC :: Ce : CB$$
$$\tfrac{1}{2} W : h :: Ce : \sqrt{\tfrac{1}{4}a^2 + h^2}$$

or,

hence,
$$Ce = \frac{\tfrac{1}{2} W \sqrt{\tfrac{1}{4}a^2 + h^2}}{h}$$

But Ce is the strain on CB. Assuming the average breaking weight of wrought-iron at 60000 lbs. per square inch, if we divide Ce by 60000 we evidently obtain the number of square inches required in CB to just break under the strain Ce. Hence,

$$\text{Cross-section of } CB = \frac{Ce}{60000} = \frac{W \sqrt{\tfrac{1}{4}a^2 + h^2}}{120000\, h}$$

Multiplying this by the length CB, we get

$$\text{Volume of } CB = \frac{W(\tfrac{1}{4}a^2 + h^2)}{120000\, h} = \frac{W}{120000} \times \frac{\tfrac{1}{4}a^2 + h^2}{h}$$

If now we find out what value of h will make this expression a minimum, we have solved our problem.

This first factor being constant may be omitted, the volume being a minimum when $\frac{\tfrac{1}{4}a^2 + h^2}{h}$ is a minimum.

Differentiating, we get

$$\frac{h \times (2\,h\,dh) - (\tfrac{1}{4}a^2 + h^2)\,dh}{h^2}$$

Putting the 1st differential coefficient equal to 0, we get

$$2h^2 - \tfrac{1}{4}a^2 - h^2 = 0$$
$$h^2 = \tfrac{1}{4}a^2$$
$$h = \tfrac{1}{2}a$$

The 2d differential coefficient is

$$\frac{a^2}{2\,h^3}$$

which is positive for $h = \tfrac{1}{2} a$. Hence this value of h makes the original function a minimum. Therefore, in the isosceles triangle $D\,B\,C$, $C\,B$ should make an angle of 45° with the vertical, or the two ties should be at right angles to each other.

If h were less than $\tfrac{1}{2} a$, the length of $C B$ would be diminished, but the strain on it would be increased, and therefore its cross-section, and the *volume* of the tie would be increased.

If h were greater than $\tfrac{1}{2} a$, the tension on $C B$ would be lessened, but its length would be so much increased that its *volume* would be greater than when $h = \tfrac{1}{2} a$.

The law for the strength of all other materials under tension being the same as for iron—that the strength varies directly as the section—we may conclude that the same angle of economy holds for them under similar circumstances.

BEST ANGLE FOR A SET OF TIES.

TO DETERMINE THE PROPER ANGLE OF INCLINATION FOR THE TIES OF A TRUSS THAT THEY MAY TRANSMIT A GIVEN WEIGHT TO THE POINTS OF SUPPORT WITH THE MINIMUM AMOUNT OF METAL.

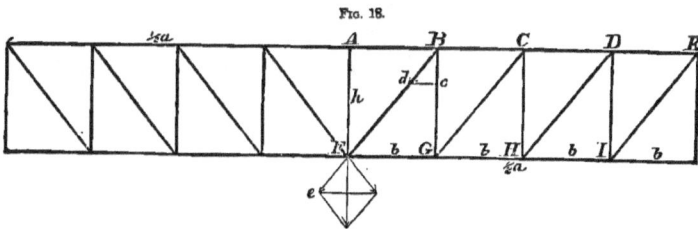

Fig. 18.

Suppose the truss drawn above, whose span is a and depth h, with a weight W suspended at the middle point F of its lower chord, which is to be transmitted to the abutments. FB, GC, HD, IE are the parallel ties which are to transmit one-half of W to the right-hand abutment. The run b of each tie is unknown, but is to be such that the sum of the volumes of the ties shall be a minimum. The angle of the posts is immaterial, except for preserving the parallelism of the ties, as it does not affect the strain on each tie, which is equal for all and independent of the inclination of the posts.

The triangles BFG and Bdc are similar. In the latter, $Bc = \frac{1}{2}W$, and in BFG, $BG = h$ and $BF = \sqrt{b^2 + h^2}$. From them we get

$$Bd : BF :: Bc : BG;$$

or,

$$Bd : \sqrt{b^2 + h^2} :: \tfrac{1}{2}W : h;$$

hence

$$Bd = \frac{W \sqrt{b^2 + h^2}}{2h}.$$

But the cross-section of BF in square inches is evidently equal to its strain divided by 60000, or $\frac{1}{60000} \times Bd$. Its length is $BF = \sqrt{b^2 + h^2}$. Hence, representing its volume by V,

$$V = \frac{W \sqrt{b^2 + h^2}}{120000\, h} \times \sqrt{b^2 + h^2} = \frac{W(b^2 + h^2)}{120000\, h}.$$

The strain and the length being the same for each tie, their volumes must be the same. The number of ties between the middle and the abutment is evidently equal to

½ a divided by b. Multiplying the volume of one tie by the number of ties, we get the entire sum which we seek. Hence,

$$\Sigma V = \frac{W(b^2 + h^2)}{120000\,h} \times \frac{a}{2b} = \frac{aW}{240000\,h} \times \frac{b^2 + h^2}{b}.$$

We wish to find such a value for b as will make this function a minimum.

The first factor is constant, and the function will be a minimum when the second factor is a minimum. Differentiating it, recollecting that h is a constant, we get

$$\frac{b \times 2b\,db - (b^2 + h^2)\,db}{b^2}.$$

Putting the first differential coefficient equal to 0, we get

$$2b^2 - (b^2 + h^2) = 0.$$
$$b^2 = h^2.$$
$$b = h.$$

The second differential coefficient is

$$\frac{2\,b\,h^2}{b^4},$$

which is positive when $b = h$. Hence this value of b makes the original function a minimum.

We see, therefore, that the angle of economy in the case of a set of ties is the same as that previously found for a pair of ties, and is 45°.

BEST ANGLE FOR A PAIR OF STRUTS.

TO DETERMINE THE PROPER ANGLE OF INCLINATION FOR A PAIR OF CAST-IRON STRUTS THAT THEY MAY TRANSMIT A GIVEN WEIGHT TO THE POINTS OF SUPPORT WITH THE MINIMUM AMOUNT OF METAL.

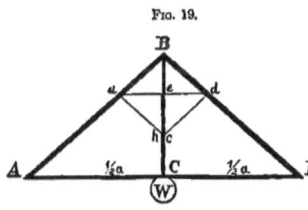

Fig. 19.

Suppose the combination $ABDC$. AB and BD are cast-iron struts, BC is a vertical tie, and AD is the chord. The weight is at C.

The weight W is at once transmitted to B, where it is Bc, and its components are Ba and Bd. The height BC is h, and the span of the truss is $AD = a$.

From Bed and BCD we get

$$Bd : BD :: Be : BC;$$

or,
$$Bd : \sqrt{h^2 + \tfrac{1}{4}a^2} :: \tfrac{1}{2}W : h;$$

whence,
$$Bd = \frac{W\sqrt{h^2 + \tfrac{1}{4}a^2}}{2h}.$$

But Bd is the strain on BD, and its length is $BD = \sqrt{h^2 + \tfrac{1}{4}a^2}$. Substituting these values in the formula for the volume of a cast-iron strut,

$$V = A \times W^{\frac{1}{1.58}} \times l^{\frac{1.79}{0.94}},$$

in which A is the numerical coefficient for either solid or hollow pillars, and writing m for $\frac{1}{1.58}$ and n for $\frac{1.79}{0.94}$, we get

$$V = A \times \left(\frac{W\sqrt{h^2 + \tfrac{1}{4}a^2}}{2h}\right)^m \times (\sqrt{h^2 + \tfrac{1}{4}a^2})^n;$$

or,
$$V = \left[A \times \left(\frac{W}{2}\right)^m\right] \times \frac{(\sqrt{h^2 + \tfrac{1}{4}a^2})^{m+n}}{h^m} = \left[A \times \left(\frac{W}{2}\right)^m\right] \frac{(h^2 + \tfrac{1}{4}a^2)^{\frac{m+n}{2}}}{h^m}.$$

In this expression for V the first factor is constant, and may be omitted in obtaining the minimum.

Differentiating, we get

$$\frac{h^m \times \frac{m+n}{2}(h^2 + \tfrac{1}{4}a^2)^{\frac{m+n}{2}-1} \times 2h\,dh - (h^2 + \tfrac{1}{4}a^2)^{\frac{m+n}{2}} \times m h^{m-1} dh}{h^{2m}}.$$

Placing the first differential coefficient equal to 0, and omitting the denominator, we get

$$h^{m+1}(m+n)(h^2 + \tfrac{1}{4}a^2)^{\frac{m+n}{2}-1} - m h^{m-1}(h^2 + \tfrac{1}{4}a^2)^{\frac{m+n}{2}} = 0$$

or, dividing by $h^{m-1}(h^2 + \frac{1}{4}a^2)^{\frac{m+n}{2}-1}$,

$$h^2(m+n) - m(h^2 + \tfrac{1}{4}a^2) = 0$$

which reduces to

$$nh^2 = \tfrac{1}{4}ma^2$$

$$h^2 = \frac{m}{n} \times \frac{a^2}{4}$$

$$h = \frac{a}{2}\sqrt{\frac{m}{n}}$$

but

$$\frac{m}{n} = 0.27933$$

and

$$\sqrt{\frac{m}{n}} = 0.528$$

Hence,

$$h = 0.528 \times \frac{a}{2} = 0.264 \times a$$

or,

$$a = 3.788 \times h.$$

This value of h substituted in the second differential coefficient of the function gives a positive result, therefore we know that it is the value which makes the function a minimum.

We see from this that the height of the triangle should be a little more than one-fourth of the span. The number 0.528 is the natural tangent of the angle of inclination of the strut with the horizontal; the angle itself is 27° 51′. The inclination with the vertical is the complement of this, or 62° 9′.

It should be borne in mind, however, that though the angle just determined is the best for economy in the struts themselves, a less angle of inclination with the vertical would diminish the tension on the chord connecting the lower extremities of the struts, and might lessen the total amount of metal required in the combination.

Assuming AD and BC to be made of wrought-iron, let us determine when the sum of their volumes is a minimum.

From Bed and BCD we get

$$de : CD :: Be : BC$$
$$de : \tfrac{1}{2}a :: \tfrac{1}{2}W : h$$

whence

$$de = \frac{aW}{4h}$$

The strain on CD is equal to de, and its length is $\tfrac{1}{2}a$. Hence its volume is

$$\frac{\frac{aW}{4h} \times \frac{1}{2}a}{5000} = \frac{a^2 W}{40000\,h}.$$

The strain on BC is W and its length is h. Hence its volume is

$$\frac{Wh}{5000}.$$

Doubling the volume of CD, for the sum of AC and CD, and adding the volume of BC, we have

$$\frac{a^2 W}{20000\,h} + \frac{Wh}{5000}$$

as the function of the variable h, whose minimum we wish to find.

The function may be written

$$\frac{W}{5000}\left(\frac{a^2}{4h} + h\right)$$

The first factor is constant and may be omitted. Differentiating the second factor, we get

$$\frac{-a^2 \times 4\,dh}{16\,h^2} + dh$$

Placing the first differential coefficient equal to 0, we get

$$\frac{-4\,a^2}{16\,h^2} + 1 = 0$$
$$16\,h^2 = 4\,a^2$$
$$h = \tfrac{1}{2}\,a$$

As this value of h makes the second differential coefficient positive, it makes the function a minimum.

The value of h previously found when the metal in the *struts* was a minimum was $0.26 \times a$, or about $\tfrac{1}{4}\,a$. Hence a height between one-quarter and one-half the span would give the greatest economy in the *total amount of metal in the combination*.

Trigonometrical Determination.

Let B be the angle CBD. Then

$$Bd = \tfrac{1}{2}\,W \sec B$$
$$BD = h \sec B$$

Substituting the value of Bd for W, and of BD for l, in the formula for the volume of a strut, and using A for the numerical coefficient, and m and n for the numerical exponents, we get

$$V = A\,(\tfrac{1}{2}\,W \sec B)^m\,(h \sec B)^n$$
or $\quad V = A\,(\tfrac{1}{2}\,W)^m \times h^n (\sec B)^{m+n}$
but $\quad h \tan B = \tfrac{1}{2}\,a$
$$h = \frac{a}{2\,\tan B}$$

Substituting this value of h, we get

$$V = [A\,(\tfrac{1}{2}\,W)^m\,(\tfrac{1}{2}\,a)^n] \times \frac{1}{(\tan B)^n} + (\sec B)^{m+n}$$

Omitting the factor in brackets, which is constant, we may write the second factor

$$\frac{\left(\frac{1}{\cos B}\right)^{m+n}}{\left(\frac{\sin B}{\cos B}\right)^{n}} = \frac{1}{(\cos B)^{m+n}} \times \frac{(\cos B)^{n}}{(\sin B)^{n}} = \frac{1}{(\cos B)^{m}(\sin B)^{n}}$$

Differentiating, we get

$$-\frac{(\sin B)^{n} \times m(\cos B)^{m-1}(-\sin B\, dB) + (\cos B)^{m} \times n(\sin B)^{n-1}(\cos B\, dB)}{(\cos B)^{2m}(\sin B)^{2n}}$$

Omitting the denominator and placing the first differential coefficient $= 0$, we get

$$(\sin B)^{n+1} \times m(\cos B)^{m-1} - (\cos B)^{m+1} \times n(\sin B)^{n-1} = 0$$
$$m \sin^{2} B = n \cos^{2} B$$
$$\tan^{2} B = \frac{n}{m} = 3.58$$
$$\tan B = 1.892$$
$$B = 62° \, 9'$$

This value of B makes the second differential coefficient positive, and therefore makes the original function a minimum. It agrees exactly with the value previously determined for the angle of inclination with the vertical.

BEST ANGLE FOR A SET OF STRUTS.

TO DETERMINE THE PROPER ANGLE OF INCLINATION FOR THE STRUTS OF A TRUSS THAT THEY MAY TRANSMIT A GIVEN WEIGHT TO THE POINTS OF SUPPORT WITH THE MINIMUM AMOUNT OF METAL.

Fig. 30.

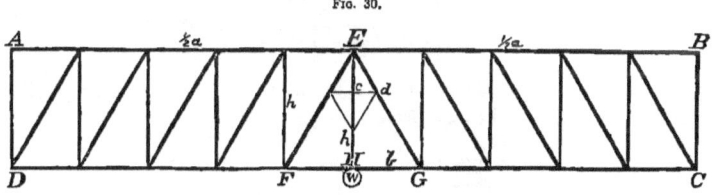

Let $ABCD$ be a truss whose depth is h, and span a. Suppose the weight W at H which is to be transmitted to C and D by the combination of inclined struts and ties. All the struts are under the same amount of compression, depending on their angle of inclination, and independent of the angle of the connecting ties. We wish to find what should be the value of b, the run of the struts, that the sum of their volumes may be a minimum. In the case of the pair of struts the value of h was to be determined—in this case h is fixed and b is the variable.

The weight W passing at once to E, where half of it is Ec, generates the strain Ed on EG. From Ecd and EHG we get

$$Ec : Ed :: EH : EG$$
$$\tfrac{1}{2} W : Ed :: h : \sqrt{b^2 + h^2}$$

Whence, $\quad Ed = \dfrac{W\sqrt{b^2 + h^2}}{2h}$

Ed is the strain on the strut EG, and $EG = \sqrt{b^2 + h^2}$ is its length. Substituting these values in the formula for the volume of a cast-iron strut, replacing the numerical coefficient by A and the exponents by m and n, we get

$$V = A \left(\frac{W\sqrt{b^2 + h^2}}{2h} \right)^m + (\sqrt{b^2 + h^2})^n$$

for the volume of EG.

The whole number of struts between H and C is evidently equal to $\tfrac{1}{2} a$ divided by b, or $\dfrac{a}{2b}$. The volumes of the struts being equal, we evidently get the sum of their volumes by multiplying the expression for EG by $\frac{a}{2b}$. Representing the sum of their volumes by ΣV, we get

$$\Sigma V = \frac{a}{2b} \times A \left(\frac{W \sqrt{b^2 + h^2}}{2h} \right)^m \times \left(\sqrt{b^2 + h^2} \right)^n$$

which is the quantity whose minimum we are seeking. The only variable in the expression is b. It may be written

$$\Sigma V = \frac{a}{2b} \times A \left(\frac{W}{2h} \right)^m \times (b^2 + h^2)^{\frac{m+n}{2}}$$

or $$\Sigma V = \left[\frac{a}{2} \times A \left(\frac{W}{2h} \right)^m \right] \times \frac{(b^2 + h^2)^{\frac{m+n}{2}}}{b}$$

As the quantity within the brackets is constant, it may be omitted in finding the minimum. Differentiating the other factor, we get.

$$\frac{b \left(\frac{m+n}{2} \right) (b^2 + h^2)^{\frac{m+n}{2} - 1} \times 2 b \, db - (b^2 + h^2)^{\frac{m+n}{2}} \times db}{b^2}$$

Making the 1st differential coefficient $= 0$, we get

$$2 b^2 \left(\frac{m+n}{2} \right) (b^2 + h^2)^{\frac{m+n}{2} - 1} = (b^2 + h^2)^{\frac{m+n}{2}}$$

whence,
$$b^2 (m + n) = b^2 + h^2$$
$$b^2 (m + n - 1) = h^2$$
$$b = \frac{h}{\sqrt{m + n - 1}}$$

Replacing m and n by their numerical values, we find that

$$\frac{1}{\sqrt{m + n - 1}} = 0.8336.$$

Hence,
$$b = h \times 0.8336.$$

Substituting this value of b in the 2d differential coefficient, we find that it gives a positive quantity; hence this value of b makes the original function a minimum.

The number 0.8336 is the natural tangent of 39° 49′. We therefore conclude that the most economical angle for the parallel struts of a set is 39° 49′ with the vertical, or 50° 11′ with the horizontal.

Trigonometrical Determination.

Let E be the angle HEG which we wish to find,

$$Ed = \tfrac{1}{2} W \sec E$$
$$EG = h \sec E$$

Substituting Ed, the strain on EG for W, and EG for l, in the formula for the volume of a strut, and using A for the numerical coefficient, and m and n for the numerical exponents, we get

$$V = A \left(\frac{W}{2} \sec E\right)^m \times (h \sec E)^n$$

Multiplying by $\frac{a}{2b}$ the number of the struts, we get

$$\Sigma V = \frac{a}{2b} \times A \left(\frac{W}{2} \sec E\right)^m \times (h \sec E)^n$$

Substituting for b its value $h \tan E$, we get

$$\Sigma V = \frac{a}{2h} \times \frac{1}{\tan E} \times A \left(\frac{W}{2}\right)^m \times h^n (\sec E)^{m+n}$$

or,
$$\Sigma V = \frac{a \times A \times W^m \times h^{n-1}}{2^{m+1}} \times \frac{(\sec E)^{m+n}}{\tan E}$$

The first factor is constant and may be omitted. Differentiating the second fraction, we get

$$\frac{\tan E \, d(\sec E)^{m+n} - (\sec E)^{m+n} \, d\tan E}{\tan^2 E}$$

or,
$$\frac{\tan E \,(m+n)\,(\sec E)^{m+n-1} \times \frac{\sin E}{\cos^2 E} dE - (\sec E)^{m+n} \times \frac{1}{\cos^2 E} dE}{\tan^2 E}$$

Putting the 1st differential coefficient $= 0$, we get

$\tan E \,(m+n)\,(\sec E)^{m+n-1} \times \sin E - (\sec E)^{m+n} = 0$
$\tan E \,(m+n) \sin E = \sec E$
$\sin^2 E \,(m+n) = 1$
$\sin^2 E = \frac{1}{m+n}$
$\log (\sin^2 E) = 1.6132924$
$\log \sin E = 9.8066462$
$E = 39° \,50' \,35''$

which agrees with our previous determination. This value of E makes the 2d differential coefficient positive, as it should.

The angle that is most economical for a single weight at the middle of a truss will evidently be the best for the general case of weights anywhere on the bridge.

LATTICE BRIDGES.

The preceding investigations are all based upon the supposition that the parts that receive and transmit the strains, are all free to act as the strains necessitate, being fastened only at their extremities. When the diagonal ties or struts are firmly connected together at their intersections, by rivets or other similar fastenings, their normal actions are hampered, and a complex resolution of strains is necessitated that cannot well be traced. This plan is always followed in lattice bridges, and is sometimes adopted for some of the bridges that we have examined. If the connections are rigid, they must interfere with the contractions or expansions of the intersecting members in directions perpendicular, or nearly so, to each other, and must bring great shearing strains on the fastenings, and possibly transverse strains on the braces.

In such cases the only safe method of discussion seems to be to consider the truss as a solid beam, with large diamond-shaped interstices, and to calculate its strength as if it had originally been solid, and had had the greater part of the metal between the chords removed. Where the intersecting braces are not thus fastened together, the discussion will be similar to that already given for whichever of the trusses examined is most like in its combination to the given lattice bridge. Similar strains will be found, varying only in amount, and we will have shorter panel lengths and more intersections of the braces. As the strains on the Top Chord are cumulative, and as the lattice bridges use the same lengths for top segments as the bridges heretofore used, they will require more metal in the Top unless each top segment is made larger towards the middle of the bridge than at the other end. They will also probably require more material on account of the extra amount of fastenings required.

Lattice bridges are often constructed without any apparent attention to the fact that some members of the web must undergo strains of compression. This may account for the lack of rigidity observable in many bridges of this class.

It is well known that pillars of any appreciable length, when under a strain of compression, break by flexure. Any arrangement that fixes the middle of such a pillar adds greatly to its strength. It practically divides it into two pillars of half the length of the original, and, as we have shown in the discussion of Fink's bridge, the single long pillar requires for the same-strength 87 per cent. more material than is needed in the two short pillars. We cannot but think that rivets, or similar fastenings at the intersection of braces, are objectionable; but any arrangements, such as sleeves, that would prevent transverse motion, and yet not interfere with the normal transmission of strains through the axes of the intersecting members, would be a great gain. Jones' bridge is the only one that can use such a device, as he alone has intersecting diagonal compression members. There is no saving of material in thus connecting intersecting tension members; the only gain would arise from the checking of the vibrations due to the passage of the moving load.

USE OF WROUGHT-IRON IN COMPRESSION MEMBERS.

In most American bridges all compression members are made of cast-iron. In English bridges they are generally made of wrought-iron. Hodgkinson investigated the subject of the use of wrought-iron to withstand compression, and his conclusions are given in the "Report of the Commissioners appointed to inquire into the Application of Iron to Railway Structures, 1849." We quote from pages 120 and 121:

"Hence we may infer that the strength of a wrought-iron pillar, whose length was such that it would yield by bending before it had sustained a pressure to injure much the material, would be as

$$\frac{D^{3.55} - d^{3.55}}{l^n}$$

where $n = 2$, in very long pillars; but in shorter ones the value of n will be reduced to any degree, or so that $n = o$, or that $l^n =$ a constant, in very short columns; showing the weakness of wrought-iron to sustain compression, and its unsuitableness for pillars."

Although we may not follow Hodgkinson so far as to say that wrought-iron should never be used for pillars where it must sustain a strain of compression, yet it seems evident that cast-iron is the better material for this purpose as it is unquestionably the cheaper. The rigidity of cast-iron is the very quality needed in a compression member, while the flexibility which makes wrought-iron so valuable for a tension member is the greatest objection to it when a compressive strain is to be met. Flexure is the very thing of all others that we must avoid in a member to resist compression, for we know from experiment that in cast-iron pillars of any length, fracture always takes place from flexure long before the crushing weight of the material is reached, and that if a pillar is so fastened that flexure is prevented or checked, its strength is much increased. The ingenuity of English engineers has been severely taxed to give wrought-iron struts such a form of section as will secure rigidity, while they refuse to use the cheaper material that has apparently been pointed out by nature.

In this country we have secured such excellent results in castings, particularly for cannon, that we feel that if ordinary care be taken in choosing our material and in making the cast, nothing can be found that will compare with cast-iron for resisting strains of compression either in reliability or in cost.

In accordance with these views, we have assumed the compression members in all the trusses examined to be made of cast-iron.

This question is not of much importance in such a comparison as we are making, as, if it should be considered desirable to use wrought-iron to resist compression, the change can as easily be made in one truss as in another. The kind of material used to sustain given strains cannot change the strains themselves except in so far as it affects the permanent weight of the structure.

CONCLUSION.

From our preceding theories and calculations, we conclude as follows, taking up the trusses in order of lightness:

The Post Truss.

Our calculations show this truss to be the lightest of the seven which we have discussed. The reason seems to be that it agrees more nearly than any other with our theoretical determination as to the most economical angles for ties and struts. The ties are at an angle of 45°, the exact angle indicated by theory, and the struts or posts are inclined. Though the latter have not the best angle to economize in their own weights, yet they probably have the best available angle to prevent waste of material in the ties, and at the same time to make the necessary connections between them, without necessitating unequal lengths in the segments of the Top. The very fact that our investigation shows that this truss, which agrees most nearly with what theory points out as the best combination, is the lightest, is an excellent corroboration of the soundness of the theories themselves. The counter-bracing in this truss is good.

The Triangular Truss.

In this truss the struts are nearer the theoretical angle than in the Post, but the ties diverge from their angle of greatest economy. Though it makes a light bridge, yet it does not appear to be well put together for stiffness, and there is no convenient method of keeping all the parts, especially those belonging to the first panel, in such continuous close contact as to prevent shock and sinking as the engine enters upon the bridge. The system of counter-bracing seems incomplete.

The Linville Truss.

This is a very good combination. The ties are at their angle of greatest economy, but there is a little loss from the fact of the posts being vertical. It seems to be necessary in trusses to transfer the weights on them to the abutments in the most rapid manner possible, and the gain in the Post truss seems in some measure due to the fact that while descending the posts the weights gain half a panel length towards the abutment.

The system of counter-bracing, when used as shown on the diagram, is all that could be desired.

The Murphy-Whipple Truss.

This is a good combination, though neither struts nor ties are placed at their angles of greatest economy, and therefore it requires more material than the trusses that precede. The system of counter-bracing is good.

The Jones Truss.

The calculations for this truss indicate that the transmission of weights by strut-braces is not so economical of material as their transmission by tie-braces. Increase of length tells more on struts than on ties, as the quantity of material in the former varies nearly as the square of the length, while in the latter it is as the first power. The counter-bracing in this truss is good.

The Fink Truss.

The calculations show that the trusses proper much exceed the trussed girders in economy. We might readily see that this truss could hardly be saving of material, as the ties are very long and the longest sustain very great tensions under very low angles, deviating considerably from the angle of greatest economy. Moreover, the method of transmitting strains is such that many weights that are ultimately to reach a particular abutment are at first sent away from it to the middle of the bridge, returning through the long and nearly horizontal principal ties. In the trusses proper, many opposing top compressions meet and neutralize each other without being transmitted to the ends of the Top Chord; but in the trussed girders all strains ultimately reach the ends of the Top, producing on all the segments the same strain as on the greatest. To this is due the great size and weight of the Top in this bridge and in Bollman's.

The panel bracing which has been added in the diagrams seems to be required to secure the necessary rigidity.

As a deck or under-grade bridge, this truss appears to the greatest advantage, particularly where the span is not great. Its facility of self-adjustment under thermometric variations is one of the points in its favor.

The Bollman Truss.

This bridge is very simple in design, but it is the least economical of all in construction. The same objections that were urged against the Fink truss hold against Bollman's, but in a greater degree. There are more long tie-braces under low angles, and all strains are carried directly to the ends of the Top, thus causing the maximum strain on every segment of the Top. Experience has shown that the panel bracing used in this truss is a necessity, and on this account we have had to insist on the same precautions in all the trusses discussed. This truss is inferior to the Fink in adjusting itself to changes of temperature.

www.ingramcontent.com/pod-product-compliance
Lightning Source LLC
Chambersburg PA
CBHW022126160426